Quantum Dots
- Fundamental and Applications

Edited by Faten Divsar

Published in London, United Kingdom

IntechOpen

Supporting open minds since 2005

Quantum Dots - Fundamental and Applications
http://dx.doi.org/10.5772/intechopen.83206
Edited by Faten Divsar

Contributors
Ramalingam Gopal, Poopathy Kathikamanagthan, Ganasen Ravi, Elangovan Thanagvel, Bojarajan Arjun
Kumar, Anatolii Isaev, Ning Tu, Dmitry Biryukov, Anatoly Zatsepin, Faten Divsar

Notice
Statements and opinions expressed in the chapters are these of the individual contributors and not
necessarily those of the editors or publisher. No responsibility is accepted for the accuracy of
information contained in the published chapters. The publisher assumes no responsibility for any
damage or injury to persons or property arising out of the use of any materials, instructions, methods
or ideas contained in the book.

First published in London, United Kingdom, 2020 by IntechOpen
IntechOpen is the global imprint of INTECHOPEN LIMITED, registered in England and Wales,
registration number: 11086078, 7th floor, 10 Lower Thames Street, London,
EC3R 6AF, United Kingdom
Printed in Croatia

British Library Cataloguing-in-Publication Data
A catalogue record for this book is available from the British Library

Additional hard and PDF copies can be obtained from orders@intechopen.com

Quantum Dots - Fundamental and Applications
Edited by Faten Divsar
p. cm.
Print ISBN 978-1-83880-918-8
Online ISBN 978-1-83880-919-5
eBook (PDF) ISBN 978-1-83880-920-1

We are IntechOpen,
the world's leading publisher of Open Access books
Built by scientists, for scientists

4,900+
Open access books available

123,000+
International authors and editors

140M+
Downloads

Our authors are among the

151
Countries delivered to

Top 1%
most cited scientists

12.2%
Contributors from top 500 universities

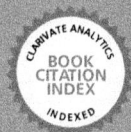

Interested in publishing with us?
Contact book.department@intechopen.com

Meet the editor

Faten Divsar received her PhD in Analytical Chemistry from Kharazmi University, Tehran, in 2009. She had postdoctoral training at Nanjing University (Nanjing, China) in 2011–2012 where her research was mainly devoted to the designing of DNA biosensors by means of electrochemiluminescence spectrometry. Currently, she is an assistant professor in the Chemistry Department of Payame Noor University, Iran. Her research focuses on two active programs, namely designing bioanalytical and electrochemical sensors and the development of innovative nanomaterial for pollution removal and water treatment.

Contents

Preface

Quantum dots (QDs) are semiconductor nanoparticles with numerous unique properties and appealing characteristics, such as size-dependent emission wavelength, narrow emission peak, and broad excitation range. A QD is an artificial molecule with energy gap and energy level spacing dependent on its size (radius) with an ability to confine electrons (quantum confinement). When the size of a QD approaches the size of the material's exciton Bohr radius, the quantum confinement effect becomes prominent and the electron energy levels can no longer be treated as a continuous band; they must be treated as discrete energy levels. It is possible to achieve highly tunable electrical and optical properties by adjusting the shape, the size, and the composition of a QD, and in doing so, facilitating the use of QDs in a broad range of applications. This book discusses a few theoretical aspects and applications of QDs. The first section substantiates new opportunities to use QDs in the developments and the study of both fundamental theories and applied physics. Precisely controlling the size, shape, emission of color, and band gap of QDs allows for their use in different applications from energy harvesting to biomedicine. The physical and chemical phenomenon of QDs can be explained through a theoretical model using quantum confinement behavior.

The second section discusses the potentials of QDs in biological imaging. It provides a comprehensive overview of QD applications in tumor targeting and cancer imaging. QDs attract great attention as contrast and therapeutic agents, owing to their unique properties of good light stability, low toxicity, and strong fluorescence intensity, and the ability to change emission wavelength with their size.

As the editor of this book, I would like to thank all the authors for their contributions and efforts in bringing up-to-date research and high-quality work to this volume.

Lastly, I gratefully acknowledge the IntechOpen publishing team for their support during the preparation of the book.

Dr. Faten Divsar
Payame Noor University,
Tehran, Iran

Section 1

Introduction

Chapter 1

Introductory Chapter: Quantum Dots

Faten Divsar

1. Introduction

Quantum dots are small regions defined in the semiconductor materials with the same size of the distance in an electron-hole pair [1]. The physics of quantum dots has been a very active and fruitful research topic. Their unique optical, photochemical, semiconductor, and catalytic properties are due to the quantum confinement. To date, chemistry, physics, and materials science have provided methods for the production of quantum dots and allow tighter control of factors affecting, for example, particle growth and size, solubility and emission properties. This book deals with the electronic and optical properties of quantum dots as an artificially fabricated device. These dots have proven to be useful systems to study a wide range of physical phenomena. These characteristics provide the potential applications of quantum dots in photovoltaic and laser devices, thin-film transistors, light-emitting diodes, and luminescent labels in biology and medicine. Some of these applications are discussed in separate sections in this book.

2. What are quantum dots?

Quantum dots are colloidal fluorescent semiconductor nanoscale crystals that were firstly produced in the early 1980s [2]. These artificial semiconductor nanoparticles typically have unique optical, electronic, and photophysical properties that make them appealing in promising applications in fluorescent biological labeling, imaging, solar cells, composites and detection and as efficient fluorescence resonance energy transfer donors [3]. Sufficiently miniaturized semiconductor particles show quantum confinement effects, which limit the energies at which electron hole pairs are present in the particles. Based on the relationship between the energy and wavelength of light (or color), the optical properties

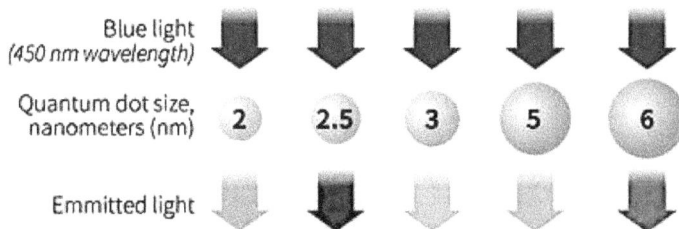

Figure 1.
Conversion of the light spectrum into different colors depends on the quantum dot size (image: RNGS Reuters/ Nanosys).

of the particle can be delicately varied depending on its size. Therefore, only by controlling the size of quantum dots, various particles with different colors can be produced to emit or absorb specific wavelengths of light [4]. As shown in **Figure 1**, different sized quantum dots emit different color light due to quantum confinement.

The physics of quantum dots counterpart with a lot of the behaviors that naturally occurring in atomic and nuclear physics of the quantum system.

3. Quantum size effects

In quantum dots, the size of particle is smaller than the Bohr radius of the electron-hole pair distance, excitons, leading to quantum confinement. During this state, the energy levels become discrete and can be predicted by the particle in a box model. The possible discrete energy levels of the electrons in such quantum dots depend on their size and accordingly to their bands gap [5]. As well, optical properties of quantum dots with sizes smaller than the de Broglie wavelength naturally depend on their electrical properties.

As shown in **Figure 2**, quantum dots as excellent luminescent materials show broad excitation range with narrow and symmetric emission spectra after excitation. Their emissions are typically much narrower than emissions of common fluorophores or organic dye molecules. In fact, the absorption of the wavelength of light with the energy equal to the band gap energy promotes the electron from the valence band to the conduction band. Afterward, the exited electron relaxes directly from the conduction band to the valence band the and emits a photon [6].

Quantum dots are becoming, nowadays, one of the fast growing and most exciting research subjects. The unique physical and optical properties of quantum dots have made them attractive tools and vectors for research in molecular biology, material science, chemical analysis, etc.

Nowadays, nanocrystals are usually prepared with atoms from groups II–VI, III–V, or IV–VI in the periodic table and eventually become many different types of quantum dots (**Table 1**).

The quantum dots can also be created by combining these systems such as GaAs–ZnS, GaAs–ZnTe, InP–ZnS, InP–ZnSe, and InP–CdS [10].

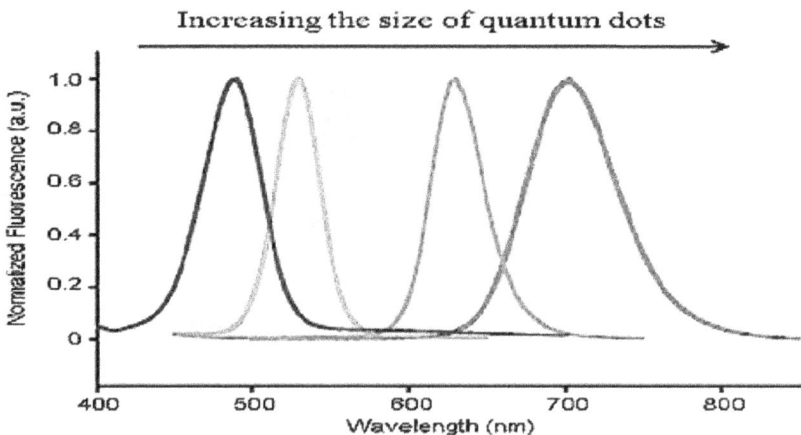

Figure 2.
Size-dependent fluorescence spectra of quantum dots.

Type	Quantum dots
IB-VI	Ag_2S, ZnAgS, CuS, $CuInS_2$, $CuInSe_2$
IIB–VI	CdS, CdSe, CdTe, ZnS, ZnSe, ZnTe, HgS, HgSe, HgTe
IIIA–V	GaAs, InGaAs, InP, InAs, InGaN
IV-VI	PbSe, PbS
IV	C, Si, Ge

Table 1.
Summary of different types of quantum dots [7–9].

4. Optical properties

Particle size is one of the most important aspects that can affect the optical properties of quantum dots. Of course, various other factors such as the shape and composition of quantum dots, as well as the methods and materials used in the synthesis of these particles also affect the optical properties and frequency of fluorescent light emitted or absorbed by these nanocrystals [11]. Therefore, by manipulating each of the effective factors, special quantum dots with specific properties can be prepared for different purposes.

5. Fabrication

Generally, there are two methods to synthesize nanomaterials including quantum dots, top-down and bottom-up (**Figure 3**). In the top-down synthesis approach, the bulk material is transferred to nanometer size using an electron beam or high-energy ions. The methods that fall into this category are electron beam lithography, reactive ion etching, focused beam lithography, and dip pen lithography [12]. However, these methods have limitations and disadvantages such as structural defects in the patterning and impurities in the quantum dots synthesized [13]. In the bottom-up methods, atoms or molecules are assembled step by step to produce nanomaterials. These synthesis approaches are highly diverse and generally

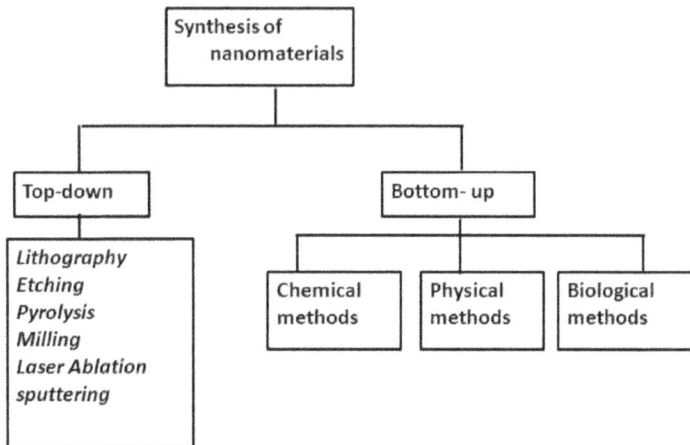

Figure 3.
Various methods for the synthesis of nanomaterials.

can be categorized as chemical, physical, and biological methods. The common bottom-up methods are derived from the development of atom or molecular self-assembly such as chemical reduction method, electrochemical method, microemulsion method, and physical/chemical vapor deposition techniques [14]. The main advantage of bottom-up approach is that they can generally synthesize homogenous nanostructures with perfect crystallographic and surface structures.

6. Potential applications

The unique optical and electrical properties of the quantum dots make them favorable for multitasking purposes. These particles emit light in visible and near-infrared region. Quantum dots are applied in electronics devices such as single electron transistors or micro-LED array, in energy applications as solar cells, photovoltaic devices, or light-emitting diodes [15]. In addition, they are also used in biological and medical sciences for imaging, labeling, detecting, and sensing. Quantum dots synthesized in organic solvents are insoluble in water. However, functionalizing the quantum dots with either hydrophilic functional groups, ligands, or by capping, organic coating can be used to transform them to aqueous soluble quantum dots. In biological applications, DNA oligonucleotide, aptamer, or antibody are grafted on quantum dots through thiol, amine, or carboxyl groups giving cross linking with molecules. Functionalized quantum dots are used for cell targeting, cell labeling, and drug delivery and in imaging. The advantages of quantum dots with respect to organic dye molecules are their high brightness, stability, and quantum efficiency [16, 17].

Author details

Faten Divsar
Department of Chemistry, Payame Noor University (PNU), Tehran, Iran

*Address all correspondence to: divsar@gmail.com

IntechOpen

References

[1] Alivisatos AP. Semiconductor clusters, nanocrystals, and quantum dots. Science. 1996;**271**:933-937

[2] Ornes S. Quantum dots. PNAS. 2016;**113**:2796-2797

[3] Pandey S, Bodas D. High-quality quantum dots for multiplexed bioimaging: A critical review. Advances in Colloid and Interface Science. 2020;**278**:102137-102139

[4] Ankireddy SR, Kim J. Synthesis of cadmium-free InP/ZnS quantum dots by microwave irradiation. Science of Advanced Materials. 2017;**9**:179-183

[5] Pattantyus-Abraham AGA, Kramer IJI, Barkhouse AR, Wang X, Konstantatos G, Debnath R, et al. Depleted-heterojunction colloidal quantum dot solar cells. ACS Nano. 2010;**4**:3374-3380

[6] Pisanic TR II, Zhang Y, Wang TH. Quantum dots in diagnostics and detection: Principles and paradigms. Analyst. 2014;**139**:2968-2981

[7] Klostranec JM, Chan WCW. Quantum dots in biological and biomedical research: Recent progress and present challenges. Advanced Materials. 2006;**18**:1953-1964

[8] Divsar F, Ju HX. Electrochemiluminescence detection of near single DNA molecules by using quantum dots–dendrimer nanocomposites for signal amplification. Chemical Communications. 2011;**47**:9879-9881

[9] Mahani M, Mousapour Z, Divsar F, Nomani A, Ju HX. A carbon dot and molecular beacon based fluorometric sensor for the cancer marker microRNA-21. Microchimica Acta. 2019;**186**:132

[10] Michael Yim W. Solid solutions in the pseudobinary (IIIV)(IIVI) systems and their optical energy gaps. Journal of Applied Physics. 1969;**40**:2617

[11] Moreels I, Lambert K, Smeets D, De Muynck D, Nollet T, Martins JC, et al. Size-dependent optical properties of colloidal PbS quantum dots. ACS Nano. 2009;**3**:3023-3030

[12] Palankar R, Medvedev N, Rong A, Delcea M. Fabrication of quantum dot microarrays using electron beam lithography for applications in analyte sensing and cellular dynamics. ACS Nano. 2013;**7**:4617-4628

[13] Valizadeh A, Mikaeili H, Samiei M, Farkhani SM, Zarghami N, Kouhi M, et al. Quantum dots: Synthesis, bioapplications, and toxicity. Nanoscale Research Letters. 2012;**7**:480

[14] Vikram P, Arpit B, Rinki G, Navin J, Jitendra P. Synthesis and applications of noble metal nanoparticles: A review. Advanced Science, Engineering and Medicine. 2017;**9**:527-544

[15] Manikandan A, Chen YZ, Shen CC, Sher CW, Kuo HC, Chueh YL. A critical review on two-dimensional quantum dots (2D QDs): From synthesis toward applications in energy and optoelectronics. Progress in Quantum Electronics. 2019;**68**:100226

[16] Wang X, Feng Y, Dong P, Huang J. A mini review on carbon quantum dots: Preparation, properties, and electrocatalytic application. Frontiers in Chemistry. 2019;**7**:1-9

[17] Cai W, Hsu AR, Li ZB, Chen X. Are quantum dots ready for in vivo imaging in human subjects? Nanoscale Research Letters. 2007;**2**:265-281

Physical Concepts of Quantum Dots

Chapter 2

Quantum Confinement Effect of 2D Nanomaterials

Gopal Ramalingam, Poopathy Kathirgamanathan,
Ganesan Ravi, Thangavel Elangovan, Bojarajan Arjun kumar,
Nadarajah Manivannan and Kaviyarasu Kasinathan

Abstract

Quantum confinement is the spatial confinement of electron–hole pairs (excitons) in one or more dimensions within a material, and also electronic energy levels are discrete. It is due to the confinement of the electronic wave function to the physical dimensions of the particles. In this effect can be divided into three ways, 1D confinement (free carrier in a plane), quantum wells; 2D confinement (carriers are free to move down), quantum wire; and 3D confinement (carriers are confined in all directions), which are discussed in detail. In addition the formation mechanism of exciton and quantum confinement behavior of strong, moderate, and weak confinement have been discussed below.

Keywords: quantum dots, energy level, exciton, confinement, Bohr radius

1. Introduction of quantum confinement

The term "quantum confinement" mainly deals with energy of confined electrons (electrons or electron hole). The energy levels of electrons will not remain continuous as in the case of bulk materials compared to the nanocrystals. Moreover, obtaining the confined electron wave functions, they become a discrete set of energy levels as shown in **Figure 1**. Such kinds of effects appear when the dimensions of the potential approach near to *de Broglie wavelength* of electrons resulting in the changes or discrete levels of energy. The effects are defined as quantum confinement and consequently, for nanocrystals, are often called quantum dots (QDs). Furthermore, this quantum dot effect has an influence in the nanomaterial properties such as electrical, optical, as well as mechanical behavior of the material. It is due to its peculiar nature why nanomaterials possess higher energy electrons than the bulk materials. Depending on the QD size, confined electrons have higher energy than the electrons in bulk materials. The semiconductor nanomaterials exhibit fascinating properties when reducing their dimensionality from 2D to 1D or 1D to 0D. Perhaps, the quantum confinement effect occurs when reducing the size and shape of nanomaterials less than 100–10 nm or even lesser. These changes due to the discrete set of electron energy levels lead to size confinement [1–3] (**Figures 1** and **2**).

In order to understand to know more about quantum confinement, it is necessary to understand the phenomenon of quantum dots (QDs). QDs are the new class of materials in which quantum confinement effects can be evident. QDs are very tiny semiconductor crystals in the order of nanometer size, and also molecules are tightly confined electrons or electron–hole pairs called "excitons" (explained in

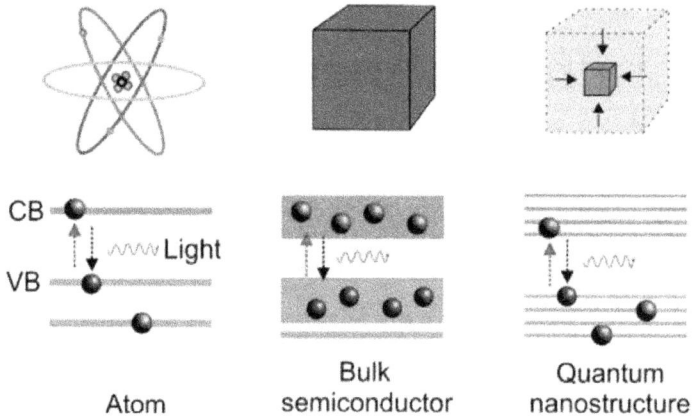

Figure 1.
Schematic diagram showing energy band structures in atom, bulk semiconductor, and quantum nanostructure.

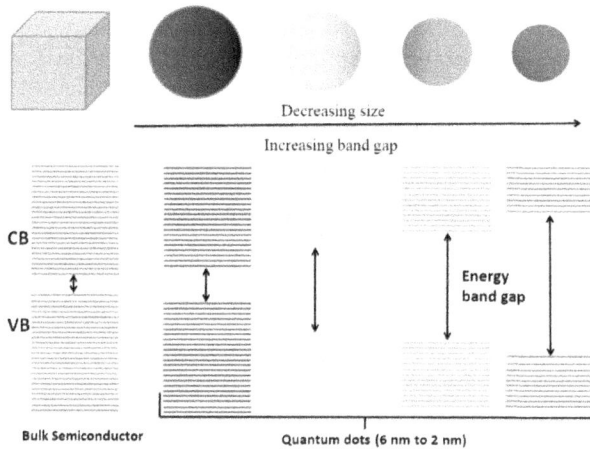

Figure 2.
Schematic diagram showing energy band structures in atom, bulk material, and quantum nanostructure.

the next section) in all three dimensions. QDs are a subatomic group in the family of nanomaterials, which comprises metals, insulators, semiconductors, and organic materials. It's well-known that the quantum confinement occurs only in semiconductor quantum dots because of their tunable bandgap nature than the zero bandgap in metals respectively. As mentioned before the peculiar tunable band gap properties QD are composed only of group II–VI-, III–V-, and IV–VI-based materials. The optical, electrical, and bandgap properties are tunable with respect to changes in particle size that lead to different multiple applications.

The main discussion is that, how bandgap can be tuned with respect to size? For that we need to understand the formation of discrete energy levels and the formation of excitons.

2. Formation of discrete energy level

To understand or recall the formation of discrete energy level, when atoms are brought together in a bulk material, the number of energy states increases substantially to form nearly continuous bands of states. And also decreasing trend occurred in the amount of atoms in the material, and energy states were delocalized with confinement nature. The phenomena create electron-hole pairs and spatially confined nature when the particles move toward the natural *de-Broglie wavelength* of electrons in the conduction band. As a result the energy difference between energy bands is increased with decreasing particle size dimension as shown in **Figure 3**.

Particle behaves like a free particle when the dimensions of the confining structure are very large in comparison to the *de Broglie wavelength*. On this stage, the energy states are continuous, and the bandgap comes to its original position, and another energy spectrum does not remain continuous and becomes discrete in nature when the dimensions of confining structure is decreased toward nanoscale. Therefore, the bandgap exhibits size-dependent properties and eventually causes a blue shift in the emitted light as the particle's size is decreased. However, this effect demonstrates the consequences of confining the electrons and electron-hole pair

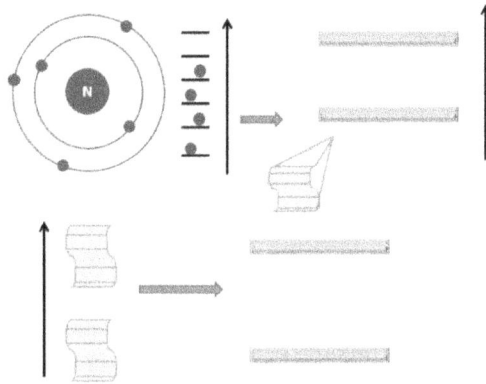

Figure 3.
Schematic diagram for the formation of discrete energy levels.

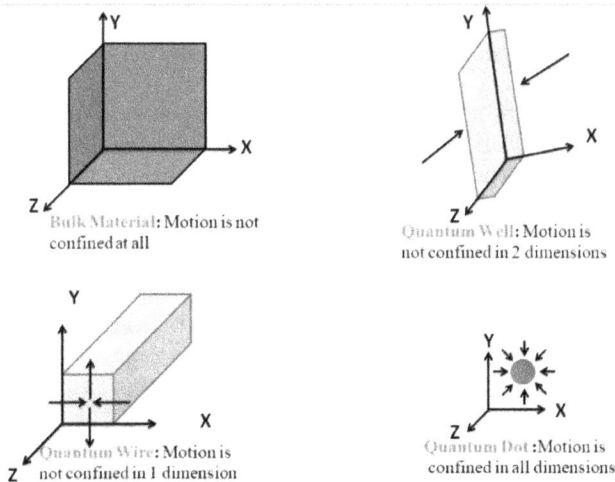

Bulk Material: Motion is not confined at all

Quantum Well: Motion is not confined in 2 dimensions

Quantum Wire: Motion is not confined in 1 dimension

Quantum Dot: Motion is confined in all dimensions

Figure 4.
Schematic representation of quantum confinement in all three directions.

(or the excitons) within a dimension which approaches the critical quantum limit, often termed as the Bohr exciton radius.

In this view, a quantum dot confines in all the three dimensions; a quantum wire (nanowire) confines in two dimensions; and a quantum well confines only in one dimension. The corresponding structures are also termed as zero-dimensional (0D), one-dimensional (1D), and two-dimensional (2D) potential wells, respectively, with regard to the number of dimensions in which the confined particle has freedom of movement. **Figure 4** shows the overview of quantum confinement in nanostructures.

- Electrons confined in one direction, i.e., **quantum wells** (thin films): Electrons can easily move in two dimensions (**2D**), so one dimensional is quantized.

- Electrons confined in two directions, i.e., **quantum wires**: Electrons can easily move in one dimension (**1D**), so two dimensional is quantized.

- Electrons confined in three directions, i.e., **quantum dots**: Electrons can easily move in zero dimension (**0D**), so three dimensional is quantized.

In conclusion, each confinement direction changes a continuous k component to a discrete component characterized by a quantum number n.

3. Formation of excitons

It is very important and necessary to understand the concept of excitons as it is the primary step to understand QDs and quantum confinement in semiconductors. In the case of semiconductor, electrons were shifted from valence band to conduction band when emitting light falling on it, and consequently recombination effect imposes or creates the photon particle. The electron and hole were occupied or created from conduction band and valence band, respectively. However, the charge of hole is equivalent to the electron charge which is helping the implementation of one particle named as excitation. In the abovementioned charges and coulomb exchange interaction, there is an attractive connection between the electron and the hole. Such kind of electron-hole pair is sometimes expressed in a simple term as quasiparticle which is named exciton. It is an electrically neutral quasiparticle that occurs in insulators, semiconductors, and some liquids.

In solid-state physics, the bandgap/energy bandgap is separated between the finite energy level of conduction and valence band. When an electron from the valence band attains sufficient energy to overcome the energy gap, due to thermal excitation or absorption of a photon, and its goes to the conduction band, a hole is created on the left behind on the valence band. The created hole is moving to conduction band; this is formed as an excited electron; the charge carrier in semiconductor device recombines with the hole after the release of energy. The combination between the electron and hole pair combination leads to the formation of excitons.

Because of the different polarity charges and the coulomb force exchange interaction, there is an attractive connection between the electron and the hole, and by a simple way, the electron-hole pair is called as a quasiparticle which is named *exciton*. Due to combination of electron and hole, the resulting neutral quasiparticle nature exits into different material natures like semiconductor, insulator, and some liquids. Furthermore this exciton transports the energy without compromising net electric charge as per condensed matter theory.

However, there is a major difference. Excitons have an average physical separation between the electron and hole, referred to as the exciton Bohr radius. This physical distance is different in each material. In the case of semiconductor nanocrystal (SNC) (QDs), the size of the particles is lesser than the Bohr radius; the electron excited by an external energy source tends to form a weak bond with its hole. This bound state of electron and electron hole, which are attracted by the electrostatic coulomb force, is often called an exciton which is shown in **Figure 5A**, **B**.

Thus, the Bohr radius is the distance in an electron–hole exciton, also called the exciton Bohr radius [4]. Every semiconductor material has a characteristic exciton Bohr radius (**Figure 5C, D**) in which the quantum confinement effect is realized. This unique confinement property causes the "band" of energies to turn into discrete energy levels in QDs (**Figure 6**).

3.1 Excitons

In general the quantum mechanical behavior of semiconductor particle, let us consider the materials in an electron which is stimulated from valence band to conduction band (**Figure 7**).

The valance band left behind electron is a 'hole', which can be thought of as a particle with its own charge (+1) and effective mass. The bound states of electron and hole are considered to be quasiparticles known or called as "exciton." Electron–hole pair formation is due to coulombic attraction. The exciton can be considered as a hydrogen-like system, and the exciton radius is easily compared with Bohr radius approximation which is used to calculate the spatial separation of the electron–hole pair:

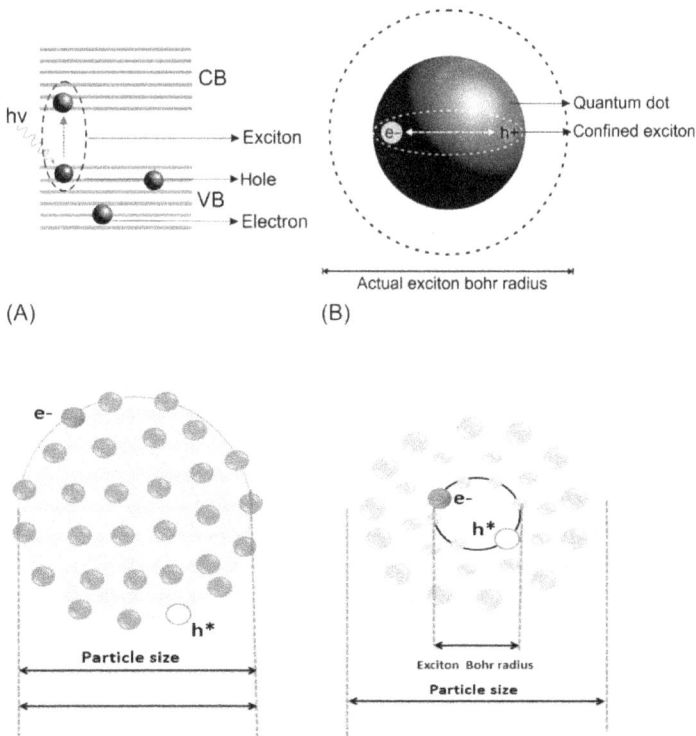

Figure 5.
(A) Formation of excitons. (B) Comparison of exciton radius and QD size.

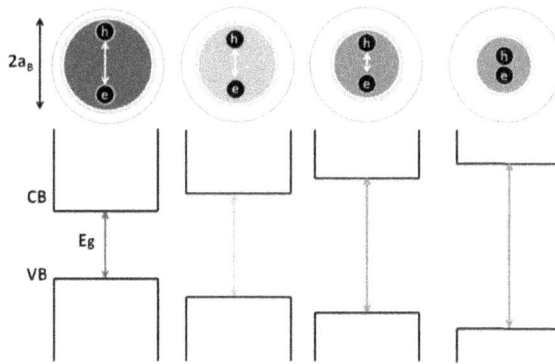

Figure 6.
Quantum confinement in semiconductor crystal.

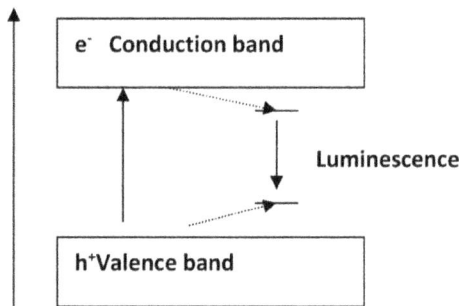

Figure 7.
Energy-level diagram showing promotion of an electron from valence band to the conduction band leaving a hole behind.

$$r = \frac{\varepsilon h^2}{\pi m_r e^2} \tag{1}$$

where r is the radius of a three-dimensional sphere containing excitation, m_r is the reduced mass of the excitation, ε is the dielectric constant of the given material, e is the charge of the electron, and h is Planck's constant. Ion cyclotron resonance is a powerful technique to find the effective mass of the hole and electron with a range of 0.1–3 m_e (m_e is the mass of the electron). Furthermore, the empirical radius and mass calculation suggest that the electron-hole pair spatial separation is around about 1–10 nm for most of the semiconductors.

4. Quantum confinement effect

In literature, semiconductor quantum dots are also known as semiconductor nanocrystals or nanoparticles. A semiconductor nanocrystal (SNC) or quantum dot (QD) is a semiconductor whose excitons are confined in all three spatial dimensions. As a result, they have properties that are between bulk semiconductors and those of discrete molecules. Size effects are observed in semiconductor crystals measuring 10–100 nm, whereas quantum size effects are usually the characteristics of nanocrystallites measuring less than 10 nm. The physicochemical properties of nanocrystalline particles are different from those of the bulk materials for two specific reasons. First, the high surface-to-volume ratio results in many atoms at the surface of the crystalline lattice. Second, the electronic bands are split up into discrete energy levels as the result of three-dimensional confinement of the charge carriers that occurs. This leads to quantum confinement of charge results in the increased bandgap with decreasing particle size.

The quantum confinement effect in low-dimensional semiconductor systems was described about 25 years ago. The bulk crystalline structure is preserved in a nanocrystal. However, due to quantum confinement, nanocrystals have molecule-like discrete electronic states which exhibit strong and size-dependent properties. In the last decade, comprehensive investigations were made to explore size-dependent properties of semiconductors with emphasis on optical properties, including absorption and luminescence.

Detail knowledge of the optical properties of the nanostructured materials is important for understanding the photophysical and photochemical process that follows the absorption of light quanta. Bawendi et al. [5] used quantum mechanical molecular orbital calculations to explain quantum confinement effect on optical absorption, accounting the well-established blue shift in the absorption spectra with decreasing particle size. Bawendi et al. [6, 7] has also provided a recent overview of this field. As discussed previous the spectral shifts during the early stages of inorganic semiconductor particle growth, molecular orbital (MO), and linear combination of atomic orbital coupled with molecular orbital (LCAO-MO) procedures provide information for the construction of energy level diagrams for clusters of several molecules up to size characteristic of bulk semiconductor.

Addition of the filled and empty orbitals of the multiple energy levels can increase the molecules level in a cluster. A decreasing trend in energy differences between the not only filled orbitals and as well as empty orbitals too. Also, a decreasing trend into the energy gap was observed between the highest occupied molecular orbitals (HOMO) and the lowest unoccupied molecular orbitals (LUMO). For a bulk semiconductor, the filled and empty states form separated continuums, i.e., the valence and conduction bands. However, for quantum dot (Q-size) regime region, energy levels were within the empty and filled states and remain discrete, and the bandgap energy levels were in higher states between the HOMO and the LUMO than that of the bulk materials as shown in the schematic diagram (**Figure 8**).

The bandgap, E_g, increases in magnitude as the semiconductor particle radius decreases in size to the point where it becomes comparable or smaller than the exciton radius [6]. These properties have led to the applications in a wide range of fields, i.e., semiconducting quantum well and super lattice devices, nonlinear optical materials, photocatalysis, and imaging systems. For II–VI compounds, maximum radii for the onset of quantum confinement effect (Q-size effects) leading to the bandgap increase have been determined by Dabbousi et al. [7] to be 10 to 100 Å. Wideband photoconductors such as TiO_2(rutile) and ZnO undergo increase in bandgap as radii approach in the range of 50 Å. It is interesting to note that a blue shift in the long-wavelength optical absorption edge with decreases in particle size was first observed by Berry [8] for AgBr samples with radii of 65 and 350 Å. It's very clear that quantum confinement effect-based semiconductor nanocrystals were produced with tremendous optical and electrical properties than that of the bulk materials which are due to the reduced behavior of characteristic length called as exciton Bohr diameter. And a characteristic length is usually in the range of few nanometers when the material reveals size-dependent optical and electrical properties.

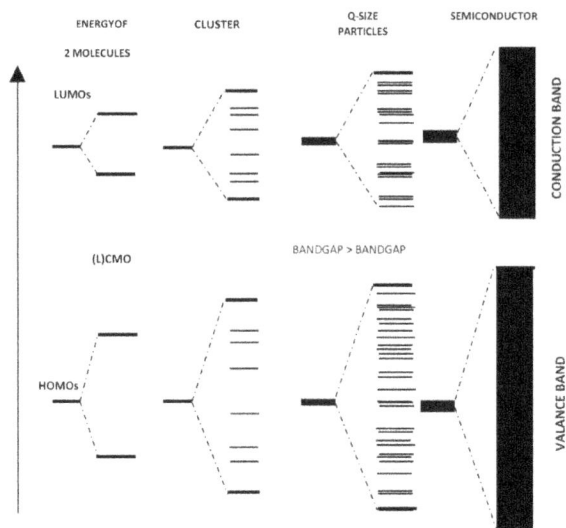

Figure 8.
A schematic diagram of the molecular orbital model for band structure.

One of the most important consequences of the spatial confinement effect is an increase in energy of the band-to-band excitation peaks (blue shift) as the radius R of a microcrystallite semiconductor is reduced in relation with the Bohr radius. However due to significant spatial confinement effect, there is an increase in energy of band-to-band excitation peaks which is called blue shift. The microcrystallite of Radius (R) ranges in semiconductor having less relation with than of the Bohr Radius. Some of minor difference would have happened between the theoretical and experimental of confine effect it is due to electron electron–hole interaction energy, in coulomb term and the confinement energy of the electron and hole in the kinetic energy.

4.1 Weak confinement regime

To observe this regime, the radius (R) of a crystallite should be greater than the bulk exciton Bohr radius (a_B). In this region of weak confinement, the dominant energy is the Coulomb term, and they already occur in size quantization of the exciton motion. The exciton energy states are shifted to higher energies by confinement, and shifts in energy ΔE are proportional to $1/R^2$. The shift "ΔE" of the exciton ground state is given approximately by

$$\Delta E \approx \frac{\hbar^2 \pi^2}{2MR^2} \tag{2}$$

where M is the mass of the exciton and it is given by $M = m_e^* + m_h^*$, with m_e^* and m_h^* being the effective masses of the electron and hole, respectively.

4.2 Moderate confinement regime

Taking another point of view of quantum confinement, especially II–VI semiconductor region, the Bohr radius is equal (a_B) to the material radius (R) which is called moderate confinement regime, and also the following term conditions should satisfy $a_h < R < a_e$ for moderate confinement regime process. A processes were observed in small QDs and a well-restricted motion of a photo-excited hole.

4.3 Strong confinement regime

Finally, the strong confinement regime was confirmed by satisfying the following condition such that $R << a_B$ and $R << a_h$. Due to these conditions, excitations are not formed, and separate size quantization of an electron and hole is the dominant factor. To strong confinement regime, must need two different reasons: the first one is the Coulomb term of electron–hole interaction is small and it's acting as a perturbation, and the second one is independent behavior of electron and holes when the above condition is applied. The optical spectra should then consist of a series of lines due to transition between sub-bands. This factor was confirmed experimentally, and the simple model gives shift in energy as a function of crystallite size as

$$\Delta E \approx \frac{\hbar^2 \pi^2}{2\mu R^2} \qquad (3)$$

in which the exciton mass M is replaced by reduced exciton mass μ, where

$$\frac{1}{\mu} = \frac{1}{m_e^*} + \frac{1}{m_h^*} \qquad (4)$$

The electrons and holes in QDs are treated as independent particles, and for the excited state, there exists a ladder of discrete energy levels as in molecular systems.

5. Summary

Recently quantum dot-based nanomaterials play a major role in the applications many field such as Q-LEDs, transistors, solar cells, laser diodes for displays, medical imaging, quantum computing, etc. In particular the QDs exhibit significant role in the optoelectronic application, its changes because of precisely controlling the size, shape, emission of color and bandgap tuning. These properties are changing inside the quantum dots, and that leads to different applications from energy harvesting to biomedical application. This entire physical and chemical phenomenon could be explained through theoretical model by using quantum confinement behavior.

Acknowledgements

The corresponding authors (Dr. G. Ramalingam & Prof. G. Ravi) acknowledge the financial support from MHRD-SPARC (ID: 890/2019), UKIERI, RUSA 2.0 grant No. F.24-51/2014-U, Policy (TN Multi-Gen) by the Government of India and UK projects.

Conflict of interest

The authors declare no conflict of interest.

Author details

Gopal Ramalingam[1*], Poopathy Kathirgamanathan[2], Ganesan Ravi[3], Thangavel Elangovan[4], Bojarajan Arjun kumar[1], Nadarajah Manivannan[5] and Kaviyarasu Kasinathan[6]

1 Department of Nanoscience and Technology, Alagappa University, Karaikudi, Tamil Nadu, India

2 Department of Chemical and Materials Engineering, Brunel University London, Uxbridge, England

3 Department of Physics, Alagappa University, Karaikudi, Tamil Nadu, India

4 Department of Energy Science, Periyar University, Salem, Tamil Nadu, India

5 Brunel University London, United Kingdom

6 UNESCO-UNISA, University of South Africa (UNISA), Pretoria, South Africa Nanosciences African network (NANOAFNET), Western Cape Province, South Africa

*Address all correspondence to: ramanloyola@gmail.com

IntechOpen

References

[1] Banyai L, Koch SW. Semiconductor quantum dots. World Scientific Series on Atomic, Molecular and Optical Physics. (Singapore) 1993;**2**:45. DOI: 10.1142/2019

[2] Murray CB, Norris DJ, Bawendi MG. Synthesis and characterization of nearly monodisperse CdE (E = sulfur, selenium, tellurium) semiconductor nanocrystallites. Journal of American Chemical Society. 1993;**115**:8706-8715

[3] Brus LE. Electron-electron and electron-hole interactions in small semiconductor. Crystallites: The size dependence of the lowest excited electronic state. The Journal of Chemical Physics. 1984;**80**:4403. DOI: 10.1063/1.447218

[4] Nozik AJ, Williams F, Nenadovic MT, Rajh T, Micic OI. Size quantization in small semiconductor particles. The Journal of Physical Chemistry. 1985;**89**:397-399. DOI: 10.1021/j100249a004

[5] Bawendi MG, Steigerwald ML, Brus LE. The quantum mechanics of larger semiconductor clusters ("quantum dots"). Annual Review of Physical Chemistry. 1990;**41**:477-496. DOI: 10.1146/annurev.pc.41.100190.002401

[6] Murray CB, Kagan CR, Bawendi MG. Self-organization of CdSe nanocrystallites into three-dimensional quantum dot superlattices. Science. 1995;**270**:1335-1338. DOI: 10.1126/science.270.5240.1335

[7] Dabbousi BO, Bawendi MG. Electroluminescence from CdSe quantum-dot/ polymer composites. Applied Physics Letters. 1995;**66**:1316. DOI: 10.1063/1.113227

[8] Berry CR. Effects of crystal surface on the optical absorption edge of AgBr. Physics Review. 1967;**153**:989. DOI: 10.1103/PhysRev.153.989

Chapter 3

Temperature Effects in the Photoluminescence of Semiconductor Quantum Dots

Anatoly Zatsepin and Dmitry Biryukov

Abstract

Temperature effects in the exciton photoluminescence specific to semiconductor quantum dots (QDs) are reviewed using Si QDs as an example. The processes of direct and indirect optical excitation of spatially confined excitons in quantum dots embedded in dielectric matrix are analyzed. The temperature behavior of the quantum dots photoluminescence (PL) excited by various methods was described in detail by a generalized electronic transitions scheme using different exciton relaxation models. The different types of temperature dependences were analyzed. The analytical expressions were obtained for their description, which allow one to determine the energy and kinetic characteristics of QD photoluminescence. It was found that the shape of the temperature dependence makes it possible to understand whether the process of exciton relaxation contains several different thermally activated stages or this is a simple one-stage process. The applicability of the obtained expressions for the analysis of the luminescence properties of quantum dots is demonstrated by the example of crystalline and amorphous silicon nanoclusters in silica matrix. It has been established that the quantum confinement effect of excitons in quantum dots leads to a decrease in the frequency characteristics and thermal activation barriers for nonradiative transitions.

Keywords: quantum dots, exciton photoluminescence, temperature dependence of luminescence, quantum confinement effects, ion implantation, mechanisms of excitation and relaxation

1. Introduction

The growing interest in the optical properties of low-dimensional systems is stimulating the development of next-generation solid-state devices in the fields of photonics, microelectronics, and optoelectronics. In particular, various technological developments use semiconductor quantum dots formed inside a dielectric matrix [1–3]. It is well known that the luminescent activity and other physical properties of such materials are largely dependent on the transformation of the electron energy spectrum of QDs caused by the size factor [1, 2]. Moreover, the photoluminescence efficiency of QDs is also affected by the mechanisms of optical excitation and the effects of quantum confinement of excitons.

According to the results of many studies [4–14], the temperature dependences of photoluminescence in dielectric and semiconductor nanostructures can significantly differ in shape and type when using various excitation methods. The PL temperature curves of quantum dots are most often presented in the form of decreasing functions with increasing temperature [4, 5, 7]. At the same time, curves with a clearly defined maximum or a monotone increase in PL intensity is sometimes observed. This is most often characteristic of direct luminescence excitation of confined excitons [9–13]. The energy transfer of emitting nanoparticles through intermediate electronic states of the matrix [4–7] can also lead to an increase in PL intensity and to curves with an extreme shape. Thus, there is a need for a detailed analysis of various energy transfer schemes and types of electronic transitions in order to explain the observed variety of forms of temperature dependences of QD photoluminescence [4–13].

It should also be said that according to some researchers, the spectral parameters of most luminescent low-scale structures are largely determined by sized and geometric factors [12, 15, 16]. In this case, the shape of the temperature curves of quantum dots PL is also substantially transformed [12]. This suggests that information on the features of the confined excitons in quantum dots can be obtained by analyzing the temperature behavior of photoluminescence. However, the lack of systematic research in this field leaves this question open.

The purpose of this chapter is to analyze the luminescence temperature dependences under direct and indirect optical excitation of spatially confined excitons in QDs. A generalized analytical description of such functional dependences is also reported.

2. Thermal and ion beam formation quantum dots

Obtaining methods of semiconductor quantum dots are questions of great importance in the fields of science and engineering. Numerous advances have been achieved in the synthesis of QDs, among them bulk etching [17], laser pyrolysis [18], gas phase synthesis [19], thermal vaporization [20], and wet chemistry techniques [21]. It should be noted that all of the above methods often require special additives such as surfactants and dopants, as well as high-temperature postprocessing, to stabilize the QDs and control their size.

Another convenient method is the thermal synthesis of quantum dots. As shown in [22–24], the synthesis of silicon QDs in the suboxide matrix takes place according to a reaction, which, depending on the temperature and duration of annealing, can be stopped at any intermediate stage: $2SiO_x \rightarrow Si\ QD + SiO_2$.

The schematic representation of the sample transformation is shown in **Figure 1**.

With ion implantation, direct accumulation of the introduced material with not only the quantum dots formation but also the quantum dots occurrence is possible due to the evolution of defective structures. In particular, there is an innovative method of creating Si quantum dots in SiO_2 under pulsed ion beam exposure [28]. In this case, quantum dots are formed as a result of the conversion and clustering of radiation defects: ODC(II) $\rightarrow$ E' $\rightarrow$ ODC(I) $\rightarrow$ Si QDs. During Gd-ion implantation with different doses, Si–O bond softening appeared, and the three main stages of defect evolution were identified: (A) formation of primary oxygen-deficient centers; (B) conversion of defects; and (C) clustering into Si QDs (**Figures** 2 and 3).

By changing the modes of radiation exposure of the SiO_2 matrix, we can control the qualitative and quantitative composition of defects and modify the optical properties of the host, including the ultraviolet transmittance and visible luminescence. As seen from **Figure 2**, the radiation defect conversion includes several consequent stages. The implementation of the described conversion mechanism provides the controlled formation of stable quantum dots at various modes of ion

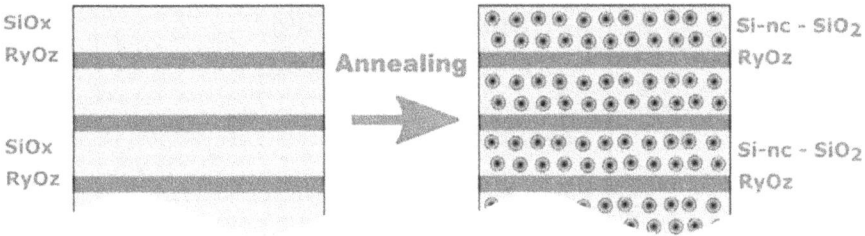

Figure 1.
Scheme of silicon nanoclusters (Si QD) formation in an initial multilayered structure under high-temperature annealing in nitrogen atmosphere. The RyOz (R—Si, Al, Zr) is a dialectical layer.

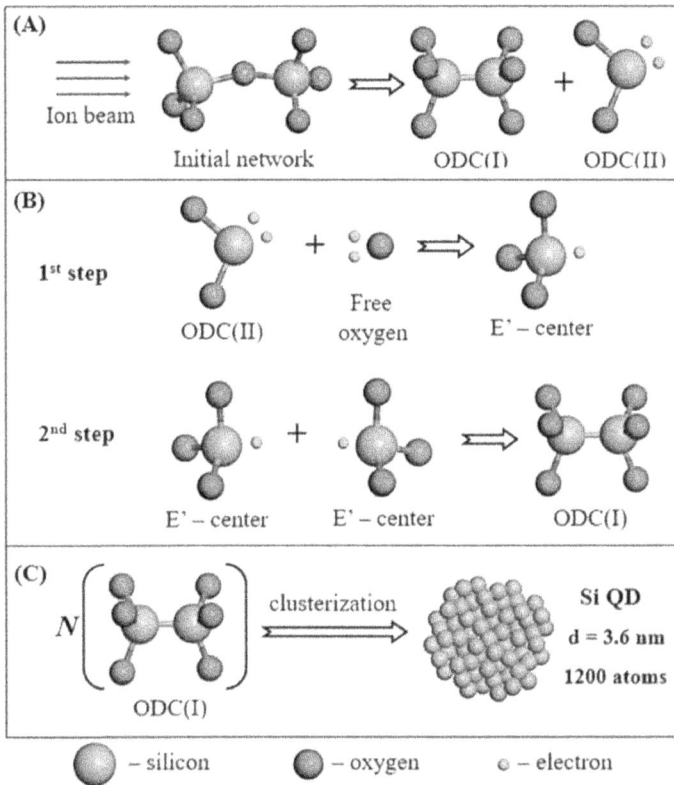

Figure 2.
Scheme of the successive stages for oxygen-deficient defect formation and fabrication of silicon quantum dots in SiO_2 subjected to ion beam irradiation. (A) Radiation formation of vacancy defects; (B) transformation of defects; (C) clusterization into quantum dots.

implantation [25]. The resulting Si QDs have a relatively small size of about 3.6 nm, which makes it possible to excite red luminescence due to the strong quantum confinement effect (**Figure 3**).

Described above the thermal synthesis of quantum dots and ion implantation methods are most convenient for stabilizing QDs in a dielectric matrix. Such systems are better suited for studying the mechanisms of excitation and radiative relaxation of QDs [26–28].

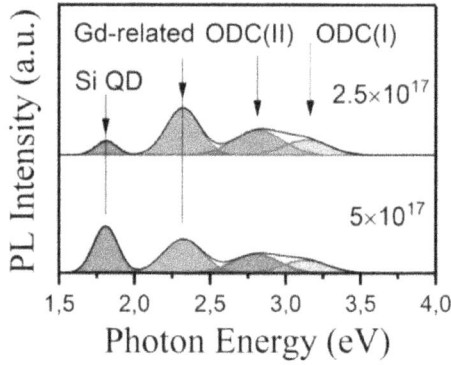

Figure 3.
Typical luminescence spectra of quantum dots and radiation defects in SiO_2 implanted with 30 keV Gd ions at fluences (2.5×10^{17} and 5×10^{17} cm^{-2}). The spectra were obtained at temperature of 8 K and excitation by photons 6.6 eV.

3. Mechanisms of QD excitation and scheme of electronic transitions

In considering the intracenter relaxation processes for spatially confined excitons in QDs, their spin state should be taken into account. Bound electron–hole pair is a diamagnetic excitation characterized by singlet and triplet levels of energy associated with mutually antiparallel and parallel spins, respectively. In accordance with the well-known Hund's rule, the triplet state is the smallest excited state because the atomic level with lower energy has a full orbital angular momentum or maximum multiplexing [29, 30].

The lifetime of triplet excitations can be several orders of magnitude longer than that of the singlet one, because triplet-singlet radiative transitions are spin-forbidden [30]. The thermally activated character of triplet luminescence after direct excitation of the singlet state is due to the energy barrier between these terms. Therefore, a growth in temperature leads to an increase in the glow intensity [10–13].

A generalized diagram for electronic transitions is shown in **Figure 4**. Direct singlet-singlet PL excitation of the QDs is shown by transition 1. At the same time, transition 6 illustrates indirect PL excitation through the electronic states of the matrix.

Each of the three different models presented in this diagram is individually characterized by a small number of fitting parameters. For better understanding of these analytic calculations, we have made the following special notations:

$\Delta E_{ISC} = (E_2–E_3)$ denotes the energy factor of the singlet-triplet intersystem crossing (ISC) relative to the QD excitons.

$E_Q = E_5$ is the activation barrier of the QD PL quenching.

$E_{ST} = E_8$ is the activation barrier of self-trapping of matrix excitons.

$\Delta E_{OC} = (E_{11}–E_{12})$ is the occupation energy factor of the radiative T_1 triplet states of the QD exciton.

$\delta P_{ISC} = p_{03}/p_{02}$ is the intersystem crossing kinetic factor for spatially confined excitons in QD.

$\delta P_T = p_{08}/P_{7(10)}$ is the energy transfer kinetic factor from the excitons of the matrix to the QDs.

$\delta P_{OC} = p_{012}/p_{011}$ is the occupation kinetic factor of the radiative T_1 triplet states of the QD exciton.

$\delta P_R = p_{05}/P_4$ is the kinetic factor for the triplet-singlet radiative transition of the QDs.

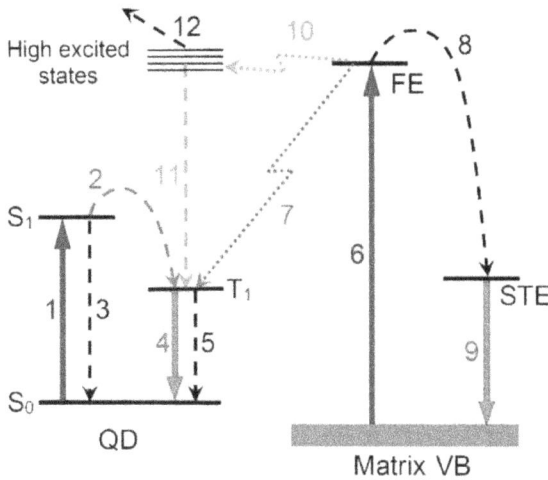

Figure 4.
Scheme of electron transitions in QD under direct and indirect excitations. Scheme shows two independent channels of excitation of the quantum dot and three different models of relaxation of excitons described by Eqs. (1)–(3). Optical and nonradiative transitions are indicated by solid and dashed arrows, respectively; the ground and excited singlet states are marked with S_o and S_1, respectively; the excited triplet state of the exciton is T_1; the levels of free and self-trapped host matrix excitons are shown as FE and STE, respectively.

Here E_2, E_3, E_5, E_8, E_{11}, and E_{12} are the thermal activation barriers. The frequency factors characterizing the nonradiative transitions 2, 3, 5, 8, 11, and 12 are designated as p_{02}, p_{03}, p_{05}, p_{08}, p_{011}, and p_{012}, respectively; the singlet-triplet radiative transition rate and energy transfer rates corresponding to transitions 7 and 10 are denoted by P_4, P_7, and P_{10}, respectively.

In the work of [10], an analytical expression is presented that well describes the extreme form of the temperature dependence of PL. It was obtained taking into account the three-level energy scheme, which includes transitions 1–5. Process with an intersystem crossing (transition 2) and a triplet-singlet luminescence (transition 4) are the main stages of the two-stage radiative recombination of the excitons. Herewith the PL intensity will be proportional to the product of the quantum efficiencies η_{ISC} and η_R for transitions 2 and 4, respectively.

Given the above and skipping the intermediate stages, the final expression for the model of QDs direct excitation can be written in the following function form:

$$I_T = I_0\eta_{ISC}\eta_R = I_0\left\{\left[1 + \delta P_{ISC}\exp\left(\frac{\Delta E_{ISC}}{kT}\right)\right]\left[1 + \delta P_R\exp\left(-\frac{E_Q}{kT}\right)\right]\right\}^{-1}, \quad (1)$$

where I_0 is the maximum luminescence intensity achieved at $\eta_{ISC} = 1$ and $\eta_R = 1$.

Eq. (1) has the four fitting parameters: δP_{ISC} and ΔE_{ISC} (transition 2) and δP_R and E_Q (transition 4). In expression (1) the physical meaning of the constants (ΔE_{ISC}, δP_{ISC}, E_Q, and δP_Q) is quite clear. The ΔE_{ISC} and δP_{ISC} parameters characterize the occupation process for T_1 triplet state, whereas E_Q and δP_Q are the parameters of the quenching process for the triplet-singlet PL. As can be seen from Eq. (1), an increase in the intensity of the triplet luminescence in the general process can be caused by the conversion of excitation from the singlet to the triplet state under some definite conditions. However, the shape of the temperature curve of the luminescence can significantly depend on various ratios between the activation energies E_2 and E_3.

Other triplet PL relaxation schemes can be considered in the same way. The energy transfer of free matrix excitons (FE) to quantum dots is confirmed experimentally [4, 5, 13]. In other words, the absorption of high-energy light quanta with the formation of free excitons (**Figure 4**, transition 6) can excite luminescence of QDs. The direct transition of the excitation to the triplet state T1 (transition 7) illustrates the simplest scheme of this process. The indicated option is similar to the direct excitation of QDs described by a two-stage process [10, 13]. The expression for the intensity of triplet luminescence (Eq. (2)) can be written assuming that transition 7 occurs without a barrier, and the formation of a self-trapped exciton (transition 8) is a competing process and also leads to luminescence (transition 9).

$$I_{T'} = I_0 \eta_T \eta_R = I_0 \left\{ \left[1 + \delta P_T \exp\left(-\frac{E_{ST}}{kT} \right) \right] \left[1 + \delta P_R \exp\left(-\frac{E_Q}{kT} \right) \right] \right\}^{-1}, \quad (2)$$

where η_R and η_T are quantum efficiencies characterizing radiative triplet-singlet transitions and energy transfer from FE to QDs, respectively. Eq. (2) has four fitting parameters: δP_R and E_Q (transition 4) and δP_T and ΔE_{ST} (transition 7), similar to Eq. (1). At the same time, the parameters (δP_R and E_Q) of nonradiative decay of excitons are independent of the excitation method, as can be seen from Eqs. (1) and (2). An analysis of the data for direct excitation of PL (**Figure 4**, transitions 4 and 5) allows one to determine these two parameters from independent measurements and then use them in the analysis of indirect excitation processes. Using this approach, one can significantly improve the procedure for approximating the experimental temperature dependences of PL and reduce the number of variable parameters.

At the same time, the large difference between the energy levels of excitons in QDs and free excitons in a dielectric matrix should be taken into account. In this regard, Eq. (2) is approximate and describes only a simplified scheme of luminescence. In fact, transition 7 cannot be considered elementary, and energy from the FE is transferred to the high-energy states (HES) of the QDs (see **Figure 4**, transition 10). Further, successive thermal transitions from HES lead to the filling of the lower triplet state T1. The indicated energy transfer sequence cannot be considered barrier-free, and triplet-singlet luminescence should be described by a multistage model.

Let us consider the generalized transition 11 with effective parameters: activation energy E_{11} and frequency factor p_{011} as a sequence of thermal transitions HES $\to \ldots \to$ T$_1$. A generalized transition 12 with the corresponding effective parameters E_{12} and p_{012} will be considered a sequence of competing processes. The energy transfer rate P_{10} to highly excited states will be assumed to be independent of temperature, and the thermal activation barrier of such a transfer is equal to zero. Then the triplet-singlet PL will be described by the following three-stage model:

$$I_{T''} = I_0 \eta_T \eta_{oc} \eta_R$$

$$= I_0 \left\{ \left[1 + \delta P_T \exp\left(-\frac{E_{ST}}{kT} \right) \right] \left[1 + \delta P_{oc} \exp\left(\frac{\Delta E_{oc}}{kT} \right) \right] \left[1 + \delta P_R \exp\left(-\frac{E_Q}{kT} \right) \right] \right\}^{-1}$$

$$(3)$$

The latter is taking into account the quantum efficiencies (η_T, η_{oc}, η_R) for energy transfer, triplet state occupation, and radiative triplet-singlet transitions, respectively.

Eq. (3) is the variant of Eq. (2), where by taking into account an additional step there are six parameters, corresponding to transition 10 (δP_T и ΔE_{ST}), transition 11 (δP_{oc} и ΔE_{oc}), and transition 4 (δP_R и E_Q). The energy transfer from the matrix to the quantum dot is denoted identically in Eqs. (2) and (3) and is described by the parameters δP_T and ΔE_{ST}. However, it should be clarified that in Eq. (2) quantum energy transfer efficiency η_T characterizes transition 7, while in Eq. (3)—transition 10.

As can be seen from **Figure 1**, radiation transition 4 with fitting parameters $(\delta P_R$ и $E_Q)$ is a common step for all considered models. Thus, the constant values of the parameters δP_R и E_Q for transitions 4 under direct or indirect excitation will significantly reduce the arbitrariness of approximation by Eqs. (1)–(3).

4. Dependence of PL temperature behavior on energy and kinetic factors

As was mentioned in previous sections, the dependence of the PL intensity on temperature $I_T(T)$ can be described by a decreasing or increasing function or a curve with an extremum. According to Eq. (1), the form of $I_T(T)$ function depends on the relations between the kinetic and activation parameters of luminescent centers. Thus, if the certain conditions determining the form of function will be known, one can associate the shape of $I_T(T)$ curve with the energy and vibrational characteristics of the emission center. In the framework of two-step PL mechanism, the shape of the curves $I_T(T)$ is determined by the multipliers $\eta_{ISC}(T)$ and $\eta_R(T)$. **Figure 5** shows the calculated functions $\eta_{ISC}(T)$ and $\eta_R(T)$ and their multiplication. The curve (3) reproduces qualitatively the temperature dependence $I_T(T)$ normalized to the number of photons incident on the surface of the sample. The next step is to discuss the influence of parameters from Eq. (1) on the form of the $I_T(T)$ dependence.

4.1 Triplet luminescence quenching

In case both temperature-dependent factors of Eq. (1) decrease with an increasing of temperature, the $I_T(T)$ function is characterized by a negative slope, wherein the quantum efficiency $\eta_R(T)$ is decreasing (see **Figure 5**, curve 1) within the range from 1 to $1/(1 + \delta P_R)$. It has a minimum which depends on the ratio p_{05}/P_4. If $\delta P_R \to 0$, then η_T will weakly depend on the temperature, keeping a value that is very close to 1 over the entire temperature range. However, the situation $\delta P_R > 1$ is most often realized, since the radiative triplet-singlet transitions are spin forbidden [30], and their rate P_4 is less than the frequency factor p_{05} for the nonradiative relaxation.

The ratio between activation barriers E_2 and E_3 determines the form of $\eta_{ISC}(T)$ function, which decreases ranging from 1 to $1/(1+\delta P_{ISC})$ if $\Delta E_{ISC} < 0$. In this case,

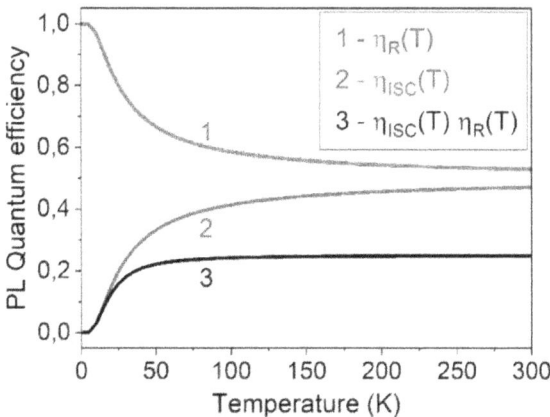

Figure 5.
Calculated temperature dependences of the quantum efficiency factors for different processes: (1) triplet-singlet radiative transitions; (2) intersystem crossing; (3) two-step process $\eta_{ISC}(T) \cdot \eta_R(T)$.

the form of $\eta_{ISC}(T)$ function is similar to that for $\eta_R(T)$, as was shown in **Figure 5**, curve 1. Thus, the PL quenching curve will have an asymptote corresponding to a certain value I_∞ (**Figure 6**, curve 1). If $\Delta E_{ISC} = 0$, the first exponent in Eq. (1) is equal to 1, which causes a temperature independence of η_{ISC} quantum efficiency. For this case, the $I_T(T)$ dependence can be described by the Mott function [31] (**Figure 6**, curve 2), and quenching of triplet PL occurs due to the nonradiative transition 5 (**Figure 4**).

The I_∞ value is determined by the parameters of the competing processes. The I_∞ value increases if δP_{ISC} and δP_R kinetic factors decrease. It should be noted that the I_∞ parameter is some hypothetical constant, because the quenching at high temperatures isn't considered in the model of direct excitation. In fact, the $I_T(T)$ dependences will tend to zero in the range of high temperatures. However, in some cases the experimental $I_T(T)$ curves can have a saturation region at room temperature [10, 12, 13]. The PL intensity in the saturation region can be contingently accepted as I_∞ value. This helps to simplify the mathematical processing.

4.2 Triplet luminescence growth

For the case of triplet PL growth, the $\eta_{ISC}(T)$ function has an increasing character because the $\eta_R(T)$ function definitely decreases. As Eq. (1) shows, the quantum efficiency of the intersystem crossing (η_{ISC}) increases with increasing of temperature if $\Delta E_{ISC} > 0$ (see **Figure 5**, curve 2), wherein the intensity of triplet PL increases up to the I_∞ value (**Figure 6**, curve 4) or the $I_T(T)$ curve has a maximum (see **Figure 6**, curve 5). If the maximum is absent, the occupation of the triplet state predominates over the process of the luminescence quenching. It is possible, when $dI_T/dT > 0$ condition is realized, which can be transformed as:

$$\eta_{ISC}(T)\delta P_{ISC}(\Delta E_{ISC}) \exp\left(\frac{\Delta E_{ISC}}{kT}\right) > \eta_R(T)\delta P_R E_Q \exp\left(-\frac{E_Q}{kT}\right) \qquad (4)$$

After some mathematical transformations. we can write:

$$n > lx + m, \qquad (5)$$

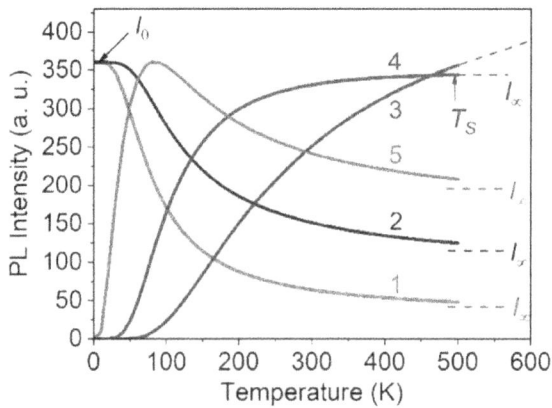

400 I_0

Figure 6.
PL temperature dependences for different ratios of energy factor ΔE_{ISC} of intersystem crossing and PL quenching thermo-activation barrier E_Q: (1) $\Delta E_{ISC} < 0$; (2) $\Delta E_{ISC} = 0$; (3) $\Delta E_{ISC} > C \cdot E_Q$; (4) $\Delta E_{ISC} = C \cdot E_Q$; (5) $0 < \Delta E_{ISC} < C \cdot E_Q$.

where

$$x = \exp\left(\frac{\Delta E_{\text{ISC}}}{kT}\right); n = \frac{E_Q}{\Delta E_{\text{ISC}}} + 1; l = \delta P_R \cdot \frac{E_Q - \Delta E_{\text{ISC}}}{\Delta E_{\text{ISC}}}; m = \frac{\delta P_R}{\delta P_{\text{ISC}}} \cdot \frac{E_Q}{\Delta E_{\text{ISC}}} \quad (6)$$

Since n > 1, finally one can obtain the following:

$$\Delta E_{\text{ISC}} > C \cdot E_Q, \quad (7)$$

where

$$C = \left(\frac{1}{\delta P_{\text{ISC}}} + 1\right)\left(\frac{1}{\delta P_R} + 1\right)^{-1} \quad (8)$$

The inequality (Eq. (7)) shows that in general the temperature dependence $I_T(T)$ is determined by both the peculiarities of the QD energy structure and their vibrational properties. From Eq. (8) one can see that the coefficient C is responsible for the relations between the kinetic factors (δP_{ISC}, δP_R) for the radiative transitions and intersystem crossing. In the case of slow kinetics of PL ($\delta P_{\text{ISC}} \gg 1$ and $\delta P_R \gg 1$) the $C \rightarrow 1$, one can neglect the vibrational properties of the QDs.

If instead of inequality (Eq. (7)) we will deal with the strict equality $\Delta E_{\text{ISC}} = C \cdot E_Q$, the $I_T(T)$ dependence will be presented as the function with saturation point at temperature $T_S \gg \Delta E_{\text{ISC}}/k$ (see **Figure 6**, curve 4).

4.3 Extremal temperature dependence of triplet luminescence

The maximum for $I_T(T)$ curve occurs under the condition that the derivative of (1) is equal to zero. Thus, the energy factor of the singlet-triplet intersystem crossing ΔE_{ISC} and the activation barrier of quenching (E_Q) have the following relationship:

$$0 < \Delta E_{\text{ISC}} < C \cdot E_Q \quad (9)$$

Previously we noted that C coefficient is introduced for accounting the influence of the rate of the vibrational processes on the form of PL temperature dependence for the QDs. However, the triplet luminescence is characterized by a relatively slow kinetics, when the C parameter is close to 1. Thus, we can state the following key points:

1. Rapidly decreasing $I_T(T)$ curves mean that the thermal activation barrier is higher than the activation energy for the intersystem crossing ($\Delta E_{\text{ISC}} < 0$).

2. In the case of temperature-independent occupation of the triplet states ($\Delta E_{\text{ISC}} = 0$), the PL quenching can be described by the Mott type function (see **Figure 6**, curve 2). The nonradiative triplet-singlet channel is responsible for this process (see **Figure 4**, transition 5).

3. If the intersystem crossing barrier is high ($\Delta E_{\text{ISC}} > C \cdot E_Q$), the intensity of the triplet luminescence increases. In this case the efficiency of intersystem crossing reaches its saturation at high temperatures.

4. The luminescence intensity increases until some point of saturation at temperature T_S when $\Delta E_{\text{ISC}} = C \cdot E_Q$. This saturation point corresponds to the balance between the quenching process and the occupation of T_1 triplet states.

5. If $0 < \Delta E_{ISC} < C \cdot E_Q$, the curve $I_T(T)$ has an extremum. In the range of $I_T(T)$, increasing the occupation process of triplet states dominates. The decreasing of PL intensity is due to the prevalence of the luminescence quenching.

6. The shape of the $I_T(T)$ dependence isn't affected by the vibrational properties of the luminescent QDs ($C \approx 1$), if the PL kinetic is relatively slow (kinetic factors δP_{ISC} and δP_R significantly greater than 1).

Analysis of Eqs. (2) and (3) can be performed in a similar way. These analytical equations describe the temperature behavior of the QD triplet luminescence at excitation within the spectral range of the exciton absorption in a wide bandgap matrix. As shown by Eq. (2), the simplified two-step model of this process corresponds to the decreasing function of the first type (see **Figure 6**, curve 1). All of the five above mentioned types of $I_T(T)$ can be described by Eq. (3) for different E_Q, E_{ST}, and ΔE_{oc} values. The main condition for increasing of this function is $\Delta E_{oc} > 0$.

Summarizing, the δP_T and ΔE_{ST} parameters characterizing the occupation of radiative states are the key parameters for the approximation of the PL temperature dependence for QDs. The values of these parameters differ for the direct and indirect excitation mechanisms. Herewith, for all three models, transition 4 is the same, so one can fix the parameters δP_R and E_Q in the approximation operation.

5. Theory and experiment comparison

5.1 A case of direct excitation

In order to check the adequacy of the model for direct excitation (Eq. (1)), we have performed the analysis of the experimental PL temperature dependences for silicon QDs with the well-known luminescence properties [4, 5, 12, 13, 31]. There are selective PL bands at 1.7–1.8 eV with a full-width maximum at half-height (FWHM) of 0.18–0.15 eV at direct excitation.

Figure 7 (curves 1 and 2) shows the experimental temperature dependences $I_T(T)$, constructed from the integrated intensities of the Si QD luminescence bands excited by the radiation 3.6 eV of nitrogen laser in the temperature range 9–300 K. To clarify the effect of the energy parameters E_{ISC} and E_Q on the shape and the luminescence intensity, we built simulated curves $I_T(T)$, which are presented in **Figure 7b**.

In **Table 1** the calculated parameters of considered PL temperature dependences (curves 3–7) are listed. **Table 1** and **Figure 7** show that increase of ΔE_{ISC} causes the reducing of PL intensity due to the reducing of the efficiency of intersystem crossing. On the contrary, increase of the activation barrier E_Q for the competing process leads to an increase in the PL intensity due to the decrease of the quenching effectiveness. For all cases, the increasing of the activation energies leads to an increasing of the Tm temperature, which corresponds to the maximum of $I_T(T)$ curve.

As was discussed in the section "Extremal temperature dependence for the triplet luminescence," all calculated curves except curve 5 belong to the fifth type of form. Curves 4 and 7 have the maxima outside the temperature range, which is shown in **Figure 7**. Curve 5 corresponds to the form of the third type. From **Table 1**, it is seen that this curve satisfies the condition (Eq. (7)).

The third-type curves demonstrate the lowest PL intensity. In the framework of developing the effective nanophosphors, the materials with the fifth-type form of

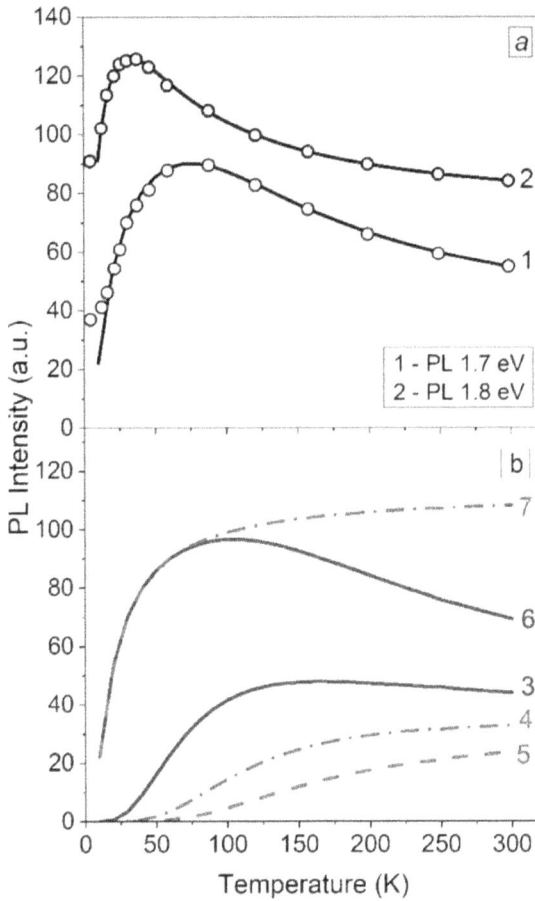

Figure 7.
PL intensity at directly excited Si QDs versus temperature in silica films implanted by Si and C ions
(a): (1) – SiO$_2$/Si (emission 1.7 eV); (2) – SiO$_2$/Si/C (emission 1.8 eV); (b) – simulated curves I$_T$(T),
obtained by using the Eq. (1).

PL temperature dependence (curves 6 and 7 in **Figure** 7) are of a great interest. Thus, both high intensity and stability of PL for a wide temperature range can be achieved if the activation energy E_Q of the PL quenching increases. The main attention was paid to the relationship between the activation energies of the nonradiative processes on the basis of the analysis of Eq. (1) and the parameters, which impact on the form of experimental PL quenching curves. However, it must not be neglected that the kinetic parameters of the emission centers also influence the shape of the temperature behavior and intensity of PL.

The PL intensity additionally increases, if the pre-exponential factors δP_{ISC} and δP_R decreases, as shown by Eq. (1). If the corresponding parameters will be tenfold different ($p_{02} > 10 \, p_{03}$ and $P_4 > 10 \, p_{05}$), this effect will be more noticeable. However, such relationships are unlikely in the case of slow kinetics of the triplet luminescence. At the same time, the PL intensity can decrease due to the influence of reducing factors, such as the ratio of the kinetic parameters. So, a wider series of real experimental dependencies should be considered and analyzed. It allows to

Curves	ΔE_{ISC}, eV	E_Q, eV	$C \cdot E_Q$, eV	T_m, K
1	0.0020	0.0260	0.0294	33
2	0.0007	0.0090	0.0090	75
3	0.0100	0.0260	0.0294	166
4	0.0200			461
5	0.0300			—
6	0.0020	0.0400	0.0452	104
7		0.2000	0.2260	378

Table 1.
*Energy and kinetic parameters of the nonradiative processes for the curves $I_T(T)$ in **Figure 7**. Comment: For strings 1 and 3–7, the correction is C = 1.13 (δP_{ISC} = 1.80, δP_R = 2.65); for string 2 C = 1 (δP_{ISC} = 1.10, δP_R = 1.09); the intensity scale parameter is Io = 317 a.u.*

determine which model parameters are more prone to changes in the characteristics of QDs and how they respond to these changes.

To solve this task, we considered the experimental results on PL of Si QDs in SiO_2 films obtained by Wang et al. [12]. In this work, authors obtained the Si QDs with different sizes by using the various doses of silicon ions. The emission bands are red-shifted from 1.65 to 1.43 eV with an increasing of ion dose. In addition, the broadening of emission bands from 0.2 to 0.4 eV was observed. However, the emission bands experienced a slight red shift (<0.1 eV) and a change in width (<0.05 eV) with an increasing of temperature.

The temperature dependences of the luminescence and their approximation by using Eq. (1) are shown in **Figure 8**. It is seen that the form of quenching curves depends on the ion fluence. On the one hand, epy curves 1 and 2 show a monotonic increase at low fluence. On the other hand, curves 3 and 4 have maxima around 158 and 163 K, respectively, at high fluence. Herewith, all temperature dependences correspond to the fifth type, as was shown by results of fitting (**Figure 8** and **Table 2**; see also "Extremal temperature dependence for the triplet luminescence" section). Maxima of curves 1 and 2 are observed outside the given temperature range. The larger ΔE_{ISC} energy factor of the intersystem crossing compared to this parameter for curves 3 and 4 explains their high-temperature position. The position of the temperature maximum is also affected by the activation barrier E_Q for the triplet-singlet PL quenching. The maximum is shifted to the high-temperature region if the barrier increases (**Table 2**). This effect is especially pronounced for the amorphous clusters.

In addition, a change in the ion fluence also impacts the kinetic factors of the exciton relaxation, which is reflected in the PL intensity. In particular, there is a significant reduction in the conversion kinetic factor δP_{ISC} (**Table 2**, pp. 3 and 4) in the initial stages of exposure, which leads to a strong increase in the PL intensity (see **Figure 8**, curves 1 and 2). At higher ion fluences, there is increasing of the radiation kinetic factor δP_R (**Table 2**, pp. 5 and 6), which results in the decreasing of the PL intensity (see **Figure 8**, curves 3 and 4), wherein the relaxation processes, which compete with the triplet PL, are characterized by a faster kinetics ($\delta P_{ISC} > 1$ and $\delta P_R > 1$). From **Table 2** one can see that the corresponding kinetic factors are varied in the values that range between 1.5 and 7.

For all curves the correction parameter C is close to 1 (**Figures 7** and **8**). It means the energy parameters have a great impact on the form of PL quenching curve. This conclusion is made on the basis of our theoretical (C = 1–1.13) evaluations and experimental results (C = 0.85–1.24) obtained in the work [12].

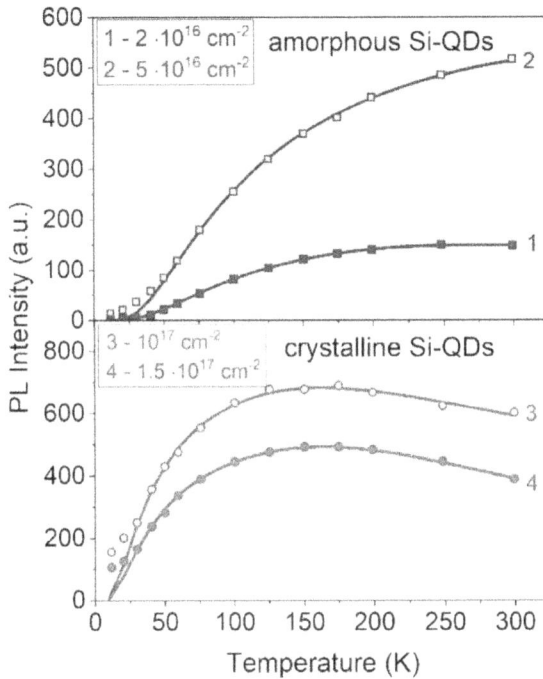

Figure 8.
PL intensity at directly excited Si QDs versus temperature in SiO$_2$/Si films implanted with different ion fluences. The circles denote an experiment [12], while the curves represent the calculated data using Eq. (1). Amorphous Si-clusters, curves (1, 2); crystalline Si-clusters, curves (3, 4). The excitation by argon laser (2.4 eV).

QD-kind	$h\nu_{PL}$, eV	δP_{ISC}	δP_R	C	ΔE_{ISC}, eV	E_Q, eV	$C \cdot E_Q$, eV	T_m, K
Amorphous	1.8	1.10	1.09	1.00	0.0007	0.0090	0.0090	33
	1.7	1.80	2.65	1.13	0.0017	0.0258	0.0292	75
	1.65	7.10	2.91	0.85	0.0100	0.0640	0.0544	271
	1.6	2.12	2.06	0.99	0.0110	0.1290	0.1277	438
Crystalline	1.5	1.50	2.80	1.23	0.0040	0.0540	0.0664	158
	1.43	2.40	7.20	1.24	0.0050	0.0710	0.0880	163

Table 2.
The approximation results of temperature dependencies for the directly excited PL in Si QDs at implanted SiO$_2$ films.

5.2 A case of indirect excitation

The low density of the QD electronic states and the high density of matrix states cause a low probability for direct excitation of QDs PL by high-energy photons (10–12 eV). By this reason, the analytical processing of PL temperature dependence for QDs should be performed with Eq. (2) or Eq. (3).

To check the models describing the indirect QD PL excitation, we have used the experimental data of QD luminescence under the synchrotron excitation (11.6 eV). The luminescence of Si QDs in amorphous SiO$_2$ films is observed at 1.8 eV (FWHM = 0.15 eV) [4, 5]. **Figure 9** shows the temperature dependences of the integrated PL intensity (circles).

The results of using Eqs. (2) and (3) for approximation are listed in **Table 3**. The OriginPro software was used for analytical processing of the obtained data and for determining the errors associated with the parameters of model. The analytical dependences of the three-stage ($I_T"(T)$) and two-stage ($I_T'(T)$) processes are shown in **Figure 9** as solid and dashed lines, respectively. The fitting error (0.3–0.8%) does not exceed the measurement error (~2%) for both cases. It means the two models are in good agreement with the experimental data for investigated temperature range. So, preferred models should be selected, taking into account experimental conditions and physical considerations regarding the energy structure of the SiO$_2$ matrix and nanoparticles.

On the basis of experimental results, the PL quenching for Si QDs starts from the liquid-helium temperature, as shown in **Figure 9**. A characteristic plateau at 100–160 K temperatures is in a good agreement with the chosen models for the indirect excitation and confirms the non-elementary mechanism of the process. The form of experimental PL quenching curve indicates the barrier-free ($E_7 = 0$) character for the transfer of the excitation energy from SiO$_2$ matrix states to Si QDs in accordance with the two-stage scheme (Eq. (2)). The decreasing dependences $I_T"(T)$ show not only the barrier-free excitation transfer ($E_{10} = 0$) but also the negative energy factor of population of T$_1$ states ($\Delta E_{oc} < 0$) in accordance with the three-stage scheme (Eq. (3)).

Transitions 10 and 11 (**Figure 4**) in the total scheme of the population process of levels T$_1$ can be ignored for the case of the thin-film SiO$_2$ matrix implanted with silicon and carbon ions, as was proven by the acceptable accuracy of the two-stage model. On the contrary, the energy factor of population of states T$_1$ (ΔE_{oc}) has a low absolute value, as was shown in **Table 3**. As noted above, this condition excludes the contribution of highly excited states to the kinetics of thermal relaxation of confined excitons. So, for the describing of PL temperature behavior for Si QDs in SiO$_2$ host at indirect excitation, the two-stage model is quite an acceptable mathematical tool.

The value of the kinetic factor (δP_T) (**Table 3**) shows that the energy transfer to the QD occurs relatively slowly. The frequency factor (p_{08}) of the exciton self-trapping is 125–150 times larger than the rate of this process. Due to the relatively

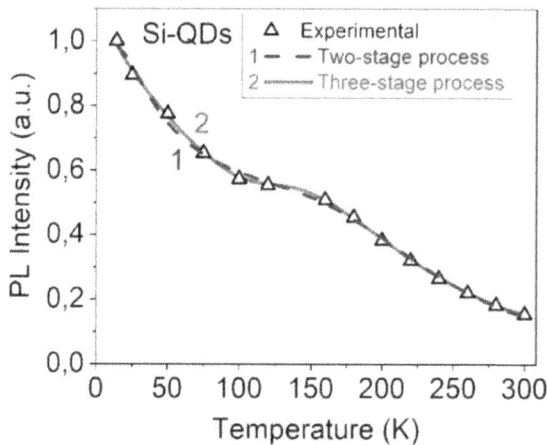

Figure 9.
PL intensity at indirectly excited amorphous Si QDs versus temperature in SiO$_2$/Si/C films. The triangles— Experimental data (emission at 1.8 eV, excitation at 11.6 eV) [4, 5]. A dashed line denotes Eq. (2) for the two-step process. The solid line represents Eq. (3) for the three-step process.

Characteristics		Two-stage model	Three-stage model
Self-trapping barrier of FE	E_{ST}, eV	0.104	0.110
Energy factor of T_1 occupation	ΔE_{OC}, eV	—	−0.002
Barrier of PL quenching	E_Q, eV	0.006	0.009
Transfer factor of kinetic energy	$\delta P_T{}^*$	125.51	153.07
Kinetic factor of T1 occupation	δP_{OC}	—	0.36
Kinetic factor for PL radiative	δP_R	1.36	1.09

*For the two-step process, δP_T corresponds to transition 7 (**Figure 4**) and for the three-step process—transition 10.

Table 3.
Results of approximation for the PL temperature dependences under direct and indirect excitations of Si QDs in SiO$_2$ films.

high activation barrier (E_{ST} = 101–110 meV) for self-trapping of the free excitons in the matrix (**Table 3**), the effective transfer of excitation FE → QDs is possible only at low temperatures. The efficiency of generation of the self-trapped excitons increases with an increasing of the temperature. As a consequence, the intensity of QD PL strongly decreases due to the additional quenching, appreciable at $T > 150$ K (**Figure 9**).

The obtained results show that the multistage relaxation of excited states at indirect excitation causes the complex character of the PL temperature dependence for QDs. So, one can ignore the participation of the highly excited states of the confined exciton and consider the two relaxation stages for the analysis of the mechanism of indirect excitation of Si QDs in the SiO$_2$ matrix [6].

In addition, the excitation method of QD luminescence also affects significantly the form of PL quenching curve. Depending on the type of excitation (direct or indirect), the form of the PL quenching curve changes dramatically. On the basis of comparison of the experimental curves (**Figures 7** and **9**), we can assume that indirect energy transfer occurs directly on the triplet state of QD by passing through the singlet state. In another case, the stage of intersystem crossing should be included in the transition scheme. Then there is also the increasing region in the PL temperature dependence at indirect excitation.

6. Dependence of the PL kinetic and energy parameters on structural and dimensional factors

In the present study, we examined the temperature dependences of the Si QD PL experimentally and observed that the energy radiative transitions are different (**Figures 7** and **8**). It's known that the spectral shift of emission band for QDs is inversely proportional to the square of its radius ($\Delta h\nu \sim R^{-2}$) [1–3, 32, 33]. On the basis of this relationship, we estimate the size of Si QDs, as listed in **Table 2**. The diameter of luminescent QDs is about 3.6–5.4 nm. This assessment is consistent with the previous literature [12].

The high-resolution transmission electron microscopy (HRTEM) shows that when the samples are implanted by Si ions with the fluences between $1.5 \cdot 10^{17}$ cm^{-2} and 10^{17} cm^{-2}, the average size of nanocrystal is between 3.8 ± 1.2 nm and 3.5 ± 1.5 nm, respectively (**Figure 10**). Based on the fact that the radius of exciton in the bulk silicon [3] is about 4.2–4.9 nm, we assume that all the samples are subjected to the effect of strong quantum confinement.

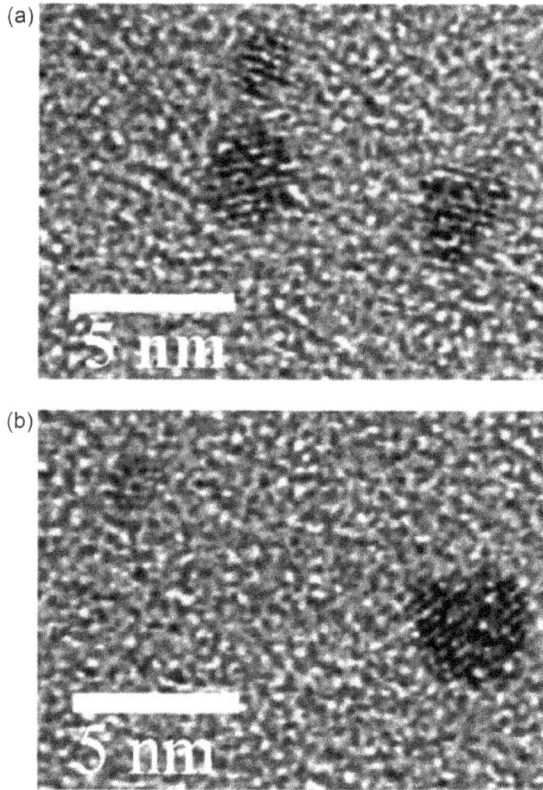

Figure 10.
HRTEM images of lattice fringe of SiO$_2$ films (250 nm), implanted by Si-ions; (a) fluence 1.5 10^{17} cm^{-2}, (b) 10^{17} cm^{-2} [12].

However, according to the diffraction data, the nanoparticles differ in size and internal structure. **Figure 11(a)** and **(b)** clearly confirms the presence of Si crystals in samples implanted by ions with fluences A and B; however, SAD does not show any sign of diffraction rings originating from anything but amorphous SiO$_2$ in **Figure 11(c)** and **(d)**. Studies have shown that the PL bands with maxima at 1.8–1.6 eV are found for the amorphous Si nanoparticles, while the emission bands at 1.5–1.43 eV are related with the crystalline Si nanoparticles [4, 5, 10, 12]. Consequently, the change in the size, morphology, and structural ordering of QDs strongly affects the dependence of PL temperature curves on the ion fluence. In addition, the disorder degree in atomic structure and position of emission bands heavily depend on the size of QDs.

The form of the PL quenching curves is determined by the activation barriers and frequency factors, which are strongly affected by the all abovementioned factors (**Figures 7** and **8**).

The analysis of physical properties of the PL of Si QDs was performed on the basis of the data listed in **Table 2**. So, when the size of QDs increases:

1. The extremum of the PL quenching curve shifts to the range of higher temperatures.

2. The value of parameter ΔE_{ISC} corresponding to the intersystem crossing of excitons increases (**Figure 12**, curve 1).

3. The activation barrier E_Q for the nonradiative relaxation increases (**Figure 12**, curve 2).

4. The kinetic factors of the intersystem crossing (δP_{ISC}) and radiation (δP_R) take a maximum value at 1.65 eV for the case of amorphous Si nanoparticles. The intersystem crossing (δP_{ISC}) and radiation (δP_R) increase for the case of crystalline Si nanoparticles.

When the structure of Si QDs is transformed from amorphous to crystalline, a dramatic reduction in the thermal activation characteristics occurs. From **Figure 12** we can see that after crystallization the transition energy parameters (ΔE_{ISC} and E_Q) are significantly reduced. Experimentally, this effect can be observed as the shift of the maximum for the $I_T(T)$ curve from 438 to 158 K (**Figure 8**).

The right interpretation of the abovementioned key points is impossible without knowledge of system of configuration curves for the emission center. The following reasons can result in the changes of the kinetic parameters and the activation barriers for nonradiative transitions:

1. The terms (configuration curves) of exciton are shifted either on the energy scale or on the coordinate configuration scale.

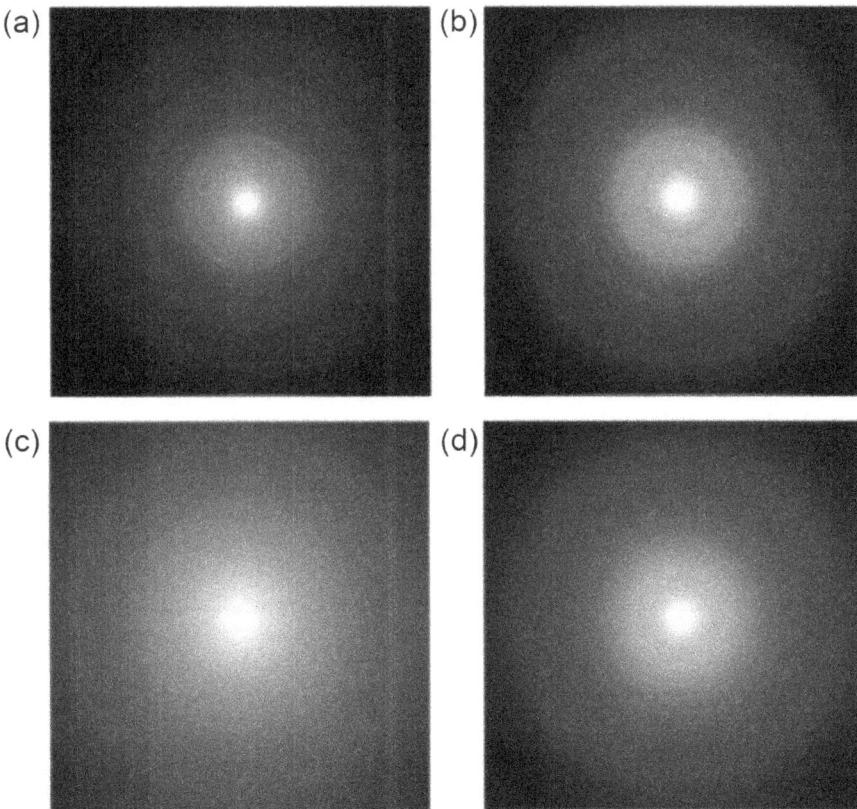

Figure 11.
Selected area diffraction (SAD) pattern images; (a) fluence 1.5×10^{17} cm^{-2}, (b) 10^{17} cm^{-2}, (c) 5×10^{16} cm^{-2}, (d) 2×10^{16} cm^{-2} [12].

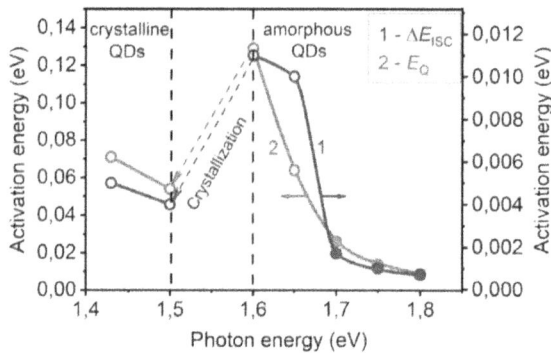

Figure 12.
The relationship between the energies of radiative PL transitions and the activation parameters of nonradiative transitions in silicon QDs in crystalline and amorphous state, which shows their dependence on structural and dimensional factors. The energy factor of the intercombining conversion is designated as E_{ISC}, and the thermally activated barrier of PL quenching is E_Q, the filled circles show the results of the data presented in [4, 5, 10, 13] and the open circles [12, 13].

2. The broadening or narrowing of configuration curves is observed.

3. The changes in the degree of continual disorder result in the distribution of configuration curves on the energy and configuration scales.

A system of hypothetical configuration curves with the terms of singlet ground state S_0 and terms of singlet and triplet excited states (S_1 and T_1, respectively) is shown in **Figure 13**. For clarity, the distribution of terms caused by the continual disorder isn't shown. It is assumed that the parameters ΔE_{ISC} and E_Q take some effective values, which correspond to the maxima of distribution functions.

The main observation of the reduction in size of nanoparticles is the increase of the optical transitions energy, owing to the shift of the terms relative to each other [1–3]. We assume that S_0, S_1, and T_1 have a parabolic form; we can see that the shift of S_1 to the position of S_1' will lead to a decrease in the energy factor ΔE_{ISC} of the intersystem crossing (see **Figure 13a**).

This is well consistent with the curve 1 of **Figure 12**. However, a similar shift of the position for T_1 toward T_1' (**Figure 13b**) implies that increasing the energy of the radiative transition ($h\nu$) will increase the activation barrier (E_Q'). At the same time, according to the experimental results (see **Figure 12**, curve 2), this barrier has to be decreased. **Figure 12** shows that this contradiction can be solved if the displacement of T_1 is accompanied by its extension (T_1''). This effect corresponds to the decrease of the frequency factor p_{05} and kinetic factor δP_R where the approximation results confirm the increasing of the energy of radiative transition (**Table 2**).

The abovementioned four key points were found experimentally [4, 5, 12, 13]. There is a red shift of emission bands from 1.8 to 1.4 eV with an increasing of QD size. In addition, the PL band broadening is observed (0.15–0.4 eV), which can be caused by the broadening of the configuration curve for the T_1 excited triplet state (**Figure 13b**).

In the case of amorphous nanoparticles (**Figure 12**), the increase of the effective parameters ΔE_{ISC} and E_Q in a larger QD size can be due to the variations in the corresponding distribution functions of energy levels. Since the size effect in the amorphous nanoparticles manifests itself more strongly in comparison with the crystalline nanoparticles, we can assume the quite importance of the continual disordering in the formation of QD optical properties. It should be noted that there

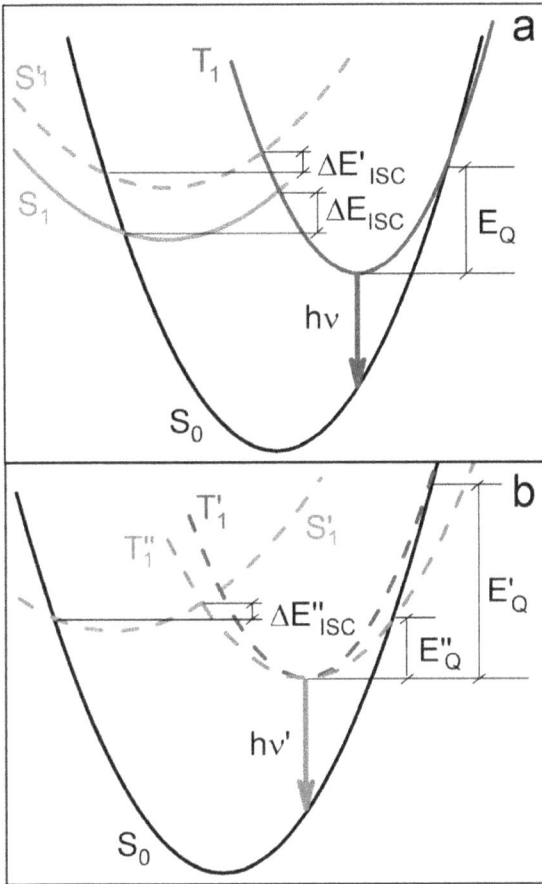

Figure 13.
Schematic illustration of the change in the energy factor ΔE_{ISC} of intercombination conversion (a) and the activation barrier E_Q of the PL quenching (b) with a decrease in the size of quantum dots. The figure shows the initial terms of the ground and excited triplet and singlet states (S_0, T_1, S_1), the displaced terms of the excited triplet and singlet states (T_1', S_1'), and the shifted and broadened term of the excited triplet state (T_1'').

is a sharp decrease in activation barriers at transition from amorphous to crystalline QDs. From this point of view, the structural disorder is becoming a negligible factor. On the other hand, the dependence of shift and broadening of the configuration curves on the QD size also are observed for the case of crystalline QDs. **Table 2** shows that the energy and the kinetic parameters of nonradiative transitions depend on the size of QDs.

The effect on structural and dimensional factors on the activation parameters of the radiative and nonradiative relaxations can be estimated from the three-level system of QD luminescence (**Figure 13**). However, for this task it is necessary to conduct special experiments and theoretical modeling (ab initio calculations). This issue is outside the scope of this work.

The developed ideas are a powerful tool for the control of the optical properties of QDs depending on the size, composition, and structure factors. We believe, the proposed approach is universal and can be applied to semiconductor QDs (such as Si, Ge, C) in dielectric hosts. The estimation of the applicability of this model to QDs of another type (e.g., with complex composition or special structural features [34]) can be performed using a special experimental check.

7. Conclusion

This chapter considers the temperature behavior of luminescent semiconductor quantum dots that have formed by thermal and ion beam methods. Analytical expressions characterizing the PL intensity of confined excitons were obtained on the basis of a generalized energy scheme for direct or indirect excitation of QDs.

Five types of temperature dependences of the luminescence of quantum dots were considered, and model parameters that affect the shape of the temperature curves were analyzed. The main variables of these models are the ratios of the rates of competing relaxation processes as well as the differences of their activation barriers. The derived expressions allow us to analyze and explain the experimentally observed luminescence temperature dependences of the most diverse form.

The models presented in this chapter were tested on the example of temperature dependences of PL silicon nanoclusters in silica matrix under different optical excitations. High informativeness and sensitivity of analytical expressions to dimensional effects and structural disordering in QDs substance are shown.

It was found that indirect excitation of Si QDs leads to a decreasing PL temperature dependence, which is due to a three-stage relaxation process. At the same time, the temperature dependences of PL upon direct excitation of silicon quantum dots are in the form of curves with a maximum and are characterized by a two-stage relaxation process. The increased effect of Si QD photoluminescence with increasing temperature, which is observed experimentally, can be explained by an increase in the density of triplet excitations acting as radiative states.

The observed structural and dimensional effects are explained using a hypothetical configurational scheme. We have found that the confined exciton effect in silicon nanoclusters manifests itself in the form of a decrease in the thermal activation barriers of nonradiative processes. Crystallization of the silicon amorphous nanoclusters also leads to a sharp decrease in the energy parameters of all thermally activated processes.

The data presented in this chapter can be used in the field of new functional materials and devices for nanophotonics. The developed ideas here can be considered as a tool for predicting and controlling the luminescent properties of quantum dots under their composition, size, and structural ordering change.

Author details

Anatoly Zatsepin and Dmitry Biryukov*
Institute of Physics and Technology, Ural Federal University, Yekaterinburg, Russia

*Address all correspondence to: a.f.zatsepin@urfu.ru; bir-70@list.ru

IntechOpen

References

[1] Zhang JZ. Optical Properties and Spectroscopy of Nanomaterials. New York: World Scientific Publishing Co Pte Ltd; 2009. p. 383

[2] Pei Y, Zeng XC. Investigating the structural evolution of thiolate protected gold clusters from first-principles. Nanoscale. 2012;**4**: 4054-4072. DOI: 10.1039/c2nr30685a

[3] Oda S, Ferry D, editors. Silicon Nanoelectronics. Taylor & Francis Group, LLC: Boca Raton; 2006. p. 310. DOI: 10.1201/9781420028645

[4] Zatsepin AF, Buntov EA, Kortov VS, Tetelbaum DI, Mikhaylov AN, Belov AI. Mechanism of quantum dot luminescence excitation within implanted SiO_2:Si:C films. Journal of Physics: Condensed Matter. 2012;**24**:045301. DOI: 10.1088/0953-8984/24/4/045301

[5] Zatsepin AF, Buntov EA. Chapter 5. Synchrotron-excited photoluminescence spectroscopy of silicon- and carbon-containing quantum dots in low dimensional SiO_2 matrices. In: Silicon-Based Nanomaterials. New York, Heidelberg, Dordrecht, London: Springer, Springer Series in Materials Science; 2013. pp. 89-117. DOI: 10.1007/978-1-4614-8169-0

[6] Pan Z, Ueda A, Xu H, Hark SK, Morgan SH, Mu R. Photoluminescence of Er-doped ZnO nanoparticle films via direct and indirect excitation. Journal of Nanophotonics. 2012;**6**:063508. DOI: 10.1117/1.JNP.6.063508

[7] Reznitsky AN, Klochikhin AA, Permogorov SA. Temperature dependence of photoluminescence intensity of self-assembled CdTe quantum dots in the ZnTe matrix under different excitation conditions. Physics of the Solid State. 2012;**54**:123-133. DOI: 10.1134/S1063783412010283

[8] Canham LT. Silicon quantum wire array fabrication by electrochemical dissolution of wafers. Applied Physics Letters. 1990;**57**:1046-1048. DOI: 10.1063/1.103561

[9] Nordin MN, Clowes J, Li SK, Curry RJ. Temperature dependent optical properties of PbS nanocrystals. Nanotechnology. 2012;**23**:275701. DOI: 10.1088/0957-4484/23/27/275701

[10] Biryukov DY, Zatsepin AF. Analytical temperature dependence of the photoluminescence of semiconductor quantum dots. Physics of the Solid State. 2014;**56**:635-638. DOI: 10.1134/S1063783414030056

[11] Nagornykh SN, Pavlenkov VI, Mikhailov AN, Belov AI, Krasil'nikova LV, Kryzhkov DI, et al. Model of photoluminescence from ion-synthesized silicon nanocrystal arrays embedded in a silicon dioxide matrix. Technical Physics. The Russian Journal of Applied Physics. 2012;**57**:1672-1675. DOI: 10.1134/S1063784212120213

[12] Wang J, Righini M, Gnoli A, Foss S, Finstad T, Serincan U, et al. Thermal activation energy of crystal and amorphous nano-silicon in SiO_2 matrix. Solid State Communications. 2008;**147**: 461-464. DOI: 10.1016/j.ssc.2008.07.011

[13] Zatsepin AF, Biryukov DY. The temperature behavior and mechanism of exciton luminescence in quantum dots. Physical Chemistry Chemical Physics. 2017;**19**:18721-18730. DOI: 10.1039/c7cp03357e

[14] Kapoor M, Singh VA, Johri GK. Origin of the anomalous temperature dependence of luminescence in semiconductor nanocrystallites. Physical Review B. 2000;**61**:1941-1945. DOI: 10.1103/PhysRevB.61.1941

[15] Glinka YD, Lin SH, Hwang LP, Chen YT, Tolk NH. Size effect in self-trapped exciton photoluminescence from SiO$_2$-based nanoscale materials. Physical Review B. 2001;**64**:085421. DOI: 10.1103/PhysRevB.64.085421

[16] Glinka YD, Lin SH, Chen YT. Time-resolved photoluminescence study of silica nanoparticles as compared to bulk type-III fused silica. Physical Review B. 2002;**66**:035404. DOI: 10.1103/PhysRevB.66.035404

[17] Kang Z, Liu Y, Tsang CHA, Ma DDD, Fan X, Wong N-B, et al. Water-soluble silicon quantum dots with wavelength-tunable photoluminescence. Advanced Materials. 2009;**21**(6):661-664. DOI: 10.1002/adma.200801642

[18] Erogbogbo F, Chang C-W, May JL, Liu L, Kumar R, Law W-C, et al. Bioconjugation of luminescent silicon quantum dots to gadolinium ions for bioimaging applications. Nanoscale. 2012;**4**:5483-5489. DOI: 10.1039/c2nr31002c

[19] Guruvenket S, Hoey JM, Anderson KJ, Frohlich MT, Krishnan R, Sivaguru J, et al. Synthesis of silicon quantum dots using cyclohexasilane Si$_6$H$_{12}$. Journal of Materials Chemistry C. 2016;**4**:8206-8213. DOI: 10.1039/c6tc01435f

[20] Roy D, Majhi K, Mondal MK, Saha SK, Sinha S, Chowdhury P. Silicon quantum dot-based fluorescent probe: Synthesis characterization and recognition of thiocyanate in human blood. ACS Omega. 2018;**3**(7):7613-7620. DOI: 10.1021/acsomega.8b00844

[21] Askari S, Macias-Montero M, Velusamy T, Maguire P, Svrcek V, Mariotti D. Silicon-based quantum dots: Synthesis, surface and composition tuning with atmospheric pressure plasmas. Journal of Physics D: Applied Physics. 2015;**48**:314002. DOI: 10.1088/0022-3727/48/31/314002

[22] Gan J, Li Q, Hu Z, Yu W, Gao K, Sun J, et al. Study on phase separation in a-SiO$_X$ for Si nanocrystal formation through the correlation of photoluminescence with structural and optical properties. Applied Surface Science. 2011;**257**(14):6145-6151. DOI: 10.1016/j.apsusc.2011.02.019

[23] Zatsepin AF, Buntov EA, Zatsepin DA, Kurmaev EZ, Pustovarov VA, Ershov AV, et al. Energy band gaps and excited states in Si QD/SiO$_X$/R$_Y$O$_Z$ (R = Si, Al, Zr) suboxide superlattices. Journal of Physics: Condensed Matter. 2019;**31**(41):415301. DOI: 10.1088/1361-648X/ab30d6

[24] Ershov AV, Chugrov IA, Tetelbaum DI, Mashin AI, Pavlov DA, Nezhdanov AV, et al. Thermal evolution of the morphology, structure, and optical properties of multilayer nanoperiodic systems produced by the vacuum evaporation of SiO and SiO$_2$. Semiconductors. 2013;**47**:481-486. DOI: 10.1134/S1063782613040064

[25] Zatsepin AF, Kuznetsova Yu A, Wong CH. Creation of Si quantum dots in a silica matrix due to conversion of radiation defects under pulsed ion-beam exposure. Physical Chemistry Chemical Physics. 2019;**21**:25467-25473. DOI: 10.1039/c9cp04715h

[26] Zatsepin AF, Zatsepin DA, Boukhvalov DW, Gavrilov NV, Shur VY, Esin AA. The MRO-accompanied modes of Re-implantation into SiO$_2$-host matrix: XPS and DFT based scenarios. Journal of Alloys and Compounds. 2017;**728**:759-766. DOI: 10.1016/j.jallcom.2017.09.036

[27] Green RJ, Zatsepin DA, Onge DJS, Kurmaev EZ, Gavrilov NV, Zatsepin AF, et al. Electronic band gap reduction and intense luminescence in Co and Mn ion-implanted SiO$_2$. Journal

of Applied Physics. 2014;**115**:103708. DOI: 10.1063/1.4868297

[28] Zatsepin AF, Pustovarov VA, Kortov VS, Buntov EA, Fitting HJ. Time-resolved photoluminescence of implanted $SiO_2:Si^+$ films. Journal of Non-Crystalline Solids. 2009;**355**: 1119-1122. DOI: 10.1016/j. jnoncrysol.2009.01.048

[29] Zatsepin AF. Statics and dynamics of excited states of oxygen-deficient centers in SiO_2. Physics of the Solid State. 2010;**52**:1176-1187. DOI: 10.1134/ S1063783410060107

[30] Sulimov VB, Sokolov VO, Dianov EM, Poumellec B. Photoinduced structural transformations in silica glass: The role of oxygen vacancies in the mechanism of formation of refractive-index gratings by UV irradiation of optical fibres. Quantum Electronics. 1996;**26**:988-993. DOI: 10.1070/ QE1996v026n11ABEH000857

[31] Mott NF, Davis EA. Electronic Processes in Non-crystalline Materials. Oxford: Oxford University Press; 1979. p. 608

[32] Vaccaro L, Spallino L, Zatsepin AF, Buntov EA, Ershov AV, Grachev DA, et al. Photoluminescence of Si nanocrystals embedded in SiO_2: Excitation/emission mapping. Physica Status Solidi B: Basic Solid State Physics. 2015;**252**:600-606. DOI: 10.1002/ pssb.201451285

[33] De Boer WDAM, Timmerman D, Dohnalová K, Yassievich IN, Zhang H, Buma WJ, et al. Red spectral shift and enhanced quantum efficiency in phonon-free photoluminescence from silicon nanocrystals. Nature Nanotechnology. 2010;**5**:878-884. DOI: 10.1038/nnano.2010.236

[34] Pan SS, Li FD, Liu QW, Xu SC, Luo YY, Li GH. Strong localization induced anomalous temperature dependence exciton emission above 300

K from SnO_2 quantum dots. Journal of Applied Physics. 2015;**117**:173101. DOI: 10.1063/1.4919595

New Opportunities to Use Quantum Dots

Chapter 4

Quantum Dots CdSe/ZnS as a Source Array of Entangled States

Anatolii Isaev

Abstract

A quantum dot is a quantum system in classical space with unique characteristics, as a result of a large quantum limitation. The experimental results of this chapter substantiate the ability of quantum dots to play a key role in purely quantum processes, for example, teleportation of quantum objects, and the generation of macroscopic quantum gravity force and, of course, are a qubit in quantum computing. A quantum dot has the ability to capture (capture) a photo-induced charge carrier by a surface defect of its crystal structure and, thereby, create a second stable long-lived quantum state, which is a necessary requirement for a qubit. This ability puts a quantum dot out of competition with respect to many other quantum objects, like qubits, in terms of the simplicity and cheapness of their continuous generation in standard laboratory conditions. Quantum dots have received wide recognition because of their unique exciton luminescence characteristics; this chapter substantiates a fundamentally new area to use quantum dots in the development and study of both fundamental and applied physics.

Keywords: quantum dots, metastable excitons, qubit, quantum entanglement

1. Introduction

Semiconductor quantum dots (QDs), for example, CdSe/ZnS is an attractive quantum object in classical space. Attractiveness is based on the unique characteristics of nanoscale structures with a large quantum limitation. The high quantum yield of exciton luminescence (up to 80%), the narrow band of this luminescence, the long photo stability and the rearrangement of the exciton luminescence band in a wide spectrum range, depending on the size of the nanoparticles, have specific unique characteristics of these crystalline nanostructures [1]. These characteristics provide potential applications in photovoltaic and laser devices, thin-film transistors, light-emitting diodes and luminescent labels in biology and medicine [2–6]. In this article, I want to justify no less, but rather more, meaningful applications, both in fundamental and in applied physics.

We are talking about the properties of crystalline structures to have on their surface quantum defects called surface trap states. These surface traps capture photo-induced charge carriers, usually an electron, and delay its recombination for a fairly long time [7–9]. In [10], the exciton luminescence of CdSe/ZnS QDs was recorded in the millisecond time range and its intensity was six orders of magnitude less than the intensity of exciton luminescence immediately after the photo excitation pulse. Such a quantum state with a long existence of an exciton is called a metastable exciton. A metastable exciton is an electron—hole pair, in which an electron is captured by a surface trap with a long lifetime [10].

The typical relaxation time of exciton luminescence is the nanosecond time range. Consequently, relaxation of all excited quantum states of QDs takes place in the nanosecond range, whereas relaxation of metastable excitons takes place in the millisecond range, which is six orders of magnitude greater than the lifetime of all other quantum states of QDs. In other words, QDs with a metastable exciton are a quasistable quantum state, and can play the role of a second stable state $|1\rangle$ of QDs, as a qubit. The first stable state of such a qubit $|0\rangle$ is QD in the ground quantum state. The irradiation of a colloid QDs is a simple and practically free way to continuously generate an array of qubits in two stable quantum states, naturally, when the energy of the optical beam quanta exceeds the QD bandgap.

Obviously, we must receive the result of any quantum process, for example, quantum computing or teleportation in the classical space, in the space where we all function. For example, a digital computer operates in its "digital space." We will need it only when it "produces" a result that is understandable to us, for example, a graph or a picture, but not as a set of numbers. QDs with a metastable exciton, as a qubits in the quantum state $|1\rangle$, have unique nonlinear optical characteristics. The fact is that the electron capture by the surface trap separates the charge carriers by a distance that coincides with the size of the QD, which are several nanometers. Such a large separation of charge carriers is the source of a very large light-induced dipole moment p of an individual QD with a metastable exciton. The dipole moment p is responsible for the value of light-induced change in the refractive index [11]. Therefore, a large value of p makes it possible to record the distribution of the concentration of individual QDs with a metastable exciton using conventional interferometry. It is this distribution of the concentration of QDs in the quantum state that is the result of the quantum process in the quantum space, which is the "quantum box" by definition of the founding fathers of quantum mechanics.

We see that quantum dots can be in two stable quantum states, which allows them to be used as a qubit in all modern quantum technologies. One of these quantum states has a significantly different classical refractive index. This property makes it possible to register individual QDs in this quantum state $|1\rangle$ by interferometry methods and, thus, to register the results of quantum processes in classical space. The experimental results of this chapter substantiate and realize this possibility, which opens up new areas for the use of quantum dots in fundamental and applied physics.

2. Teleportation of CdSe/ZnS QDs in the classical space

2.1 Quantum entanglement

Quantum entanglement is a new resource of quantum physics, the same as, for example, energy [12]. New resources make it possible to discover new potentials and implement fundamentally new processes of quantum physics. The modern representation of quantum states is based on the statement that, in quantum mechanics, any physical system is described completely by a state vector $|\Psi\rangle$ in the Hilbert space H. A system with a two-dimensional Hilbert space is called a qubit (quantum bit). For several Hilbert spaces, for example, for H_A and H_B, the complete Hilbert space is the tensor product of the subsystem spaces: $H_{AB} = H_A \otimes H_B$. Any quantum system that is described by a single state vector is a pure state. Next, the density operator and other mathematical transformations are introduced. Everything, all Physics is over! Mathematics on paper remained only!

A real qubit is a quantum object that has two stable quantum states, which, as a rule, have different easily measured classical characteristics. For example, a quantum of light in the "*o*" and "*e*" states of polarization, or a neutral atom and a charged ion of this atom. And these characteristics are measured easily using classic devices. The modern representation of quantum entanglement is based on the modern representation of quantum states. A quantum state $|\Psi\rangle$ is entangled if it cannot be written as a tensor product, i.e., $|\Phi\rangle \neq |a\rangle \otimes |b\rangle$. And, here, if the quantum state is written as the sum of tensor products, then this state is entanglement.

$$|\Phi\rangle = \frac{1}{N}(|a_1\rangle \otimes |b_1\rangle + |a_2\rangle \otimes |b_2\rangle) \qquad (1)$$

As an example, a type of state is usually given

$$|\Phi\rangle = \frac{1}{\sqrt{2}}(|01\rangle - |10\rangle) \qquad (2)$$

The term "entanglement" was introduced by Schrödinger for the first time in 1935 [13]. Schrödinger introduced this term to describe the specific relationship between quantum systems, which have correlations between their dynamic quantities: position and momentum. And this relationship is expressed in an infinite set of dynamic values of two particles. Thus, Schrödinger justified one of the key properties of quantum entanglement—the complete uncertainty of the values of classical dynamic quantities such as the position and momentum of a particle. And what do we see in Eq. (2)? Here, the quantum state $|\Phi\rangle$ of a quantum superposition of one object in two basic states is written, $|0\rangle$ and $|1\rangle$ decoherence of which will give an equiprobable (1/2) result to find this object, both in the state $|01\rangle$ and in the state $|10\rangle$. Where is the uncertainty here? Complete uncertainty reflects the form of writing a quantum state $|\Psi\rangle$ quantum superposition.

$$|\Psi\rangle = \alpha|0\rangle + \beta|1\rangle, \qquad (3)$$

where α and β are complex numbers, the sum of squares of which $|\alpha|^2 + |\beta|^2 = 1$. In essence, $|\alpha|^2$ and $|\beta|^2$ are probabilities of obtaining a state of $|0\rangle$ or $|1\rangle$ in classical space as a result of decoherence. To feel that there is entanglement you can simply imagine the coin that was thrown up, and it falls and rotates. While the coin is rotate, it is impossible to say what condition it is in. The coin is in a completely indefinite state, but it is exact in some of the states $|0\rangle$ or $|1\rangle$. The coin fell to the ground and here it can be said, determined or measured in which particular state and in which particular place on the earth. Some people write in their articles that a quantum object in a state of quantum superposition is simultaneously in all its basic states. But this is nonsense. A quantum object cannot be simultaneously in its two states, but to be in an uncertain state, there are no problems here.

Thus, we conclude that the quantum state $|\Psi\rangle$ in the entry form (3) is a quantum superposition, and it is an entangled quantum state. The principle of quantum superposition states that the linear combination of quantum states of all quantum objects of the participants of this superposition is also a quantum state. The linear combination provides an exponential growth of quantum states of quantum objects with two basic states (qubit) with a linear increase in the number of these qubits. This means that N quits provide 2^N entangled quantum states of quantum superposition. And if we return to physics, this means that the wave function of the state $|\Psi\rangle$ contains 2^N entangled wave functions. If N qubits occupy a macroscopic volume, then the wave function of state $|\Psi\rangle$ is macroscopic. This is a fundamental conclusion, since the macroscopic wave function is the basis of all quantum processes as a result of Bose-Einstein condensation. The way to create a Bose-Einstein condensate regardless of temperature opens up fantastic prospects for practical devices based on quantum effects. This is one of the physical resources of quantum entangled states.

2.2 Experimental implementation of multi-particle quantum superposition

Obviously, practical applications make sense with an array of entangled quantum states, the source of which is quantum superposition. Qbits are quantum objects in two basic states, the dynamic characteristics of which, for example, their location, can be easily measured in classical space. It is these qubits, more precisely, their quantum states $|0\rangle$ and $|1\rangle$ that form the quantum state of many-particle quantum superposition $|\Psi\rangle$ in the self-assembly mode. Self-assembly is a typical process of quantum physics, a typical example is Bose-Einstein condensate. Another typical example is the self-assembly of nanoscale quantum objects with a large quantum confinement [14]. Practical devices require qubits cheap and easily accessible. In addition, such qubits must function under normal conditions: they do not require ultrahigh vacuum or ultralow temperatures.

The semiconductor quantum dots (QDs) of CdSe/ZnS were used in this work as such qubits. The modern concept of quantum entanglement asserts that quantum entanglement is a consequence of some nonlocality of quantum mechanics, which cannot be explained from the standpoint of classical physics [15]. This concept is the basis for research on quantum communications, quantum cryptography and quantum networks. Let us leave the question of nonlocality "for later," and let's discuss the obvious property of quantum entanglement, which is it's decoherence. An array of tossed and rotating coins will fall to the ground. Each coin will fall on one of its sides. This particular side of the coin is the result of the interaction of all the coins, both among themselves and with external and internal forces, as they rotate. Decoherence of quantum superposition unravels all entangled quantum states into concrete quantum states $|0\rangle$ and $|1\rangle$ of each qubit in classical space and, thus, makes it possible to record the result of the interaction of forces in quantum superposition or quantum entanglement.

The dynamic principle of quantum superposition states that the quantum state of quantum superposition can occur again after decoherence, if conditions for this exist. Therefore, the continuous functioning of the states of quantum superposition according to the scheme "self-assembly of quantum superposition—decoherence under the influence of external and internal forces—self-assembly of quantum superposition again—decoherence again, etc." can occur only with continuous generation of the qubit. The quantum state of the qubit $|0\rangle$ is the ground unexcited state, which does not require external influence for its existence. The quantum state of the qubit $|1\rangle$ is a QD with a metastable exciton. Therefore, the continuous generation of this state is a necessary condition for the continuous functioning of the quantum state of quantum superposition. An optical beam with quantum energy greater than the bandgap is the driving force that is able to generate the state $|1\rangle$ continuously.

An optical beam with a wavelength of $\lambda = 437$ nm was used for this in experiments. A CW-laser was used as a source of this beam with a power of 30 mW. The scheme and methodology of the experiment are presented in detail in [16]. Here, we will focus on key phenomena that characterize quantum entanglement as a truly new resource with fundamentally different possibilities of practical application. In short, the experiment consisted in observing and registering the trace profile of an optical beam that spread through a suspension of CdSe/ZnS quantum dots. The fact is that the pattern of the beam trace profile is a pattern of wave aberrations of a light-induced lens [17, 18], which occurs in a QDs suspension, as a result of the self-action of an optical beam, which generates a $|1\rangle$ quantum state with a different refractive index compared to the refractive index quantum state $|0\rangle$. The dynamic pattern of wave aberrations of a light-induced lens reflects the dynamics of the space-time redistribution of the wave surface of a light-induced change in the refractive index. The wave surface of the light-induced refractive index is the space-time distribution of the concentration of QDs in the quantum state $|1\rangle$. Thus, registration of the redistribution of the refractive index makes it possible to measure the redistribution of the concentration of the quantum state $|1\rangle$ (QD with a metastable exciton), including as a result of the presence of such QDs in the quantum state of quantum superposition. This is the "highlight" of the experiment. The fact is that decoherence of an array of entangled quantum states in quantum space, which is a "quantum box," as defined by the founding fathers of quantum mechanics, occurs under the influence of all forces, internal and external. Including those forces, the existence of which we do not know. Thereby, the registration of the result of decoherence of quantum superposition makes it possible to detect these forces and understand their physical nature.

2.3 Experimental implementation of teleporting CT CdSe/ZnS

All the experimental results were obtained in a simple experiment, the scheme of which is shown in **Figure 1a**. This graphic also shows a typical beam trace profile pattern on a remote screen. **Figure 1b** shows a typical transformation of the pattern of the beam trace profile after the start of illumination. Here, the time of 0 ms is the beginning of the illumination of the QDs suspension, and the intensity distribution is the input beam profile without a cuvette with a suspension with QDs in the beam. The input beam parameters were: beam convergence angle $\theta = 5.45 \ 10^{-3}$; $w_0 = \lambda/\pi\theta = 28 \ \mu m$; $I_0 = 2P_l/\pi w^2 = 2436 \ W/cm^2$; $z_0 = \pi w_0^2/\lambda = 5.2 = 5.2$ mm. The thickness of the cuvette with colloid was 5 mm.

Quantum teleportation is the concept of quantum physics, which is being studied in a large number of recent published works. The main research topics are quantum communication, quantum computing and quantum networks. The term teleportation means the process by which bodies and objects are transferred from one place to another without moving along any path. The "quantum teleportation" boom begins with article [19], in which an unknown quantum state is first measured and then reconstructed at a remote place. The implementation of this information protocol requires a classical communication channel [19], and quantum entanglement [12]. The conceptual basis of such a quantum teleportation is the assertion that two quantum particles in an entangled state have some non-locality so that changes in the state of one particle immediately correlate with changes in the remote system regardless of the signal passing time between them [15]. If this concept is accepted as a physical reality, then one should assume the existence of some otherworldly forces, which, and only they, provide such a speculative correlation between remote quantum objects. This article substantiates another concept of quantum teleportation, which is really a physical reality, since this concept is the result of an experiment.

The meaning of the experiment was to look the transformation of the pattern of the beam trace profile when moving the cuvette with the QDs colloid on a rough surface, as a result of which, the QDs colloid was subjected to micro-shaking. **Figure 2** contains information about how the dimensions of the pattern of the beam trace profile change in the process of establishing a steady state and after the beginning of the movement of the cuvette with the colloid QDs. These data were obtained at the position of the cuvette along the axis $Z = -15z_0$. Here D_{hor}, R_{dw} R_{up} is the horizontal diameter, the radius of the lower half and the radius of the upper half of the pattern of the beam trace profile. The time τ is the characteristic time of exponential relaxation of processes that control the pattern of the beam trace profile during the accumulation of QDs and the establishment of a steady state. The beginning of the movement of the cuvette with colloid took place after 3 seconds of illumination.

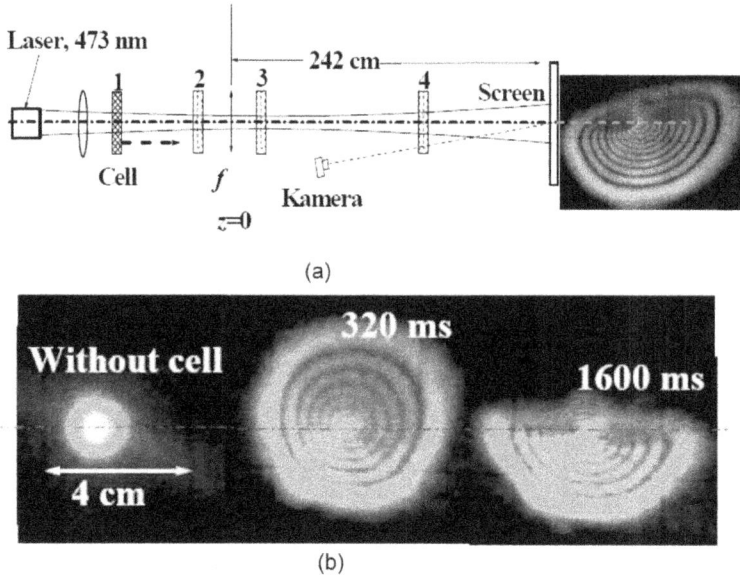

(a)

(b)

Figure 1.
(a) The scheme of the experiment and the profile of the laser beam trace on the screen. (b) The trace profile of the input optical beam and the trace profiles of the output beam in the process of accumulation of long-lived QDs after the start of illumination, z = −29z$_0$.

It is obvious that the establishment of a stationary state takes place as a result of at least two processes. The first ~400 ms there is an increase in all sizes of the pattern of the beam trace profile. Then, we see a dramatic change in the size behavior of this pattern. An obvious reduction in all sizes of this pattern is observed. We should note that the increase and subsequent reduction in the size of the pattern is well extrapolated by exponential functions. Moreover, the pattern of the upper half of the beam trace profile is reduced to a much greater degree and significantly sooner. We will analyze these experimental results below. Here, we will consider the situation after the beginning of the movement of the cuvette with the colloid to another location along the Z axis. Individual frames of the pattern transformation are shown in **Figure 3**. The real transformation of the pattern in real time is in the video files "trans1-trans3."

The beginning of this movement took place after 3 seconds of continuous illumination. Obviously, the steady state was achieved during this time (see **Figure 2**). This movement caused a complete "whistleblower" or "orgy" of the dimensions of the beam trace profile pattern, which **Figures 2** and **3** demonstrate quite well. We must note that the pattern of the profile of a beam trace changes its structure in an abrupt manner. Details of the pattern of each frame in **Figure 3** do not coincide with the details of the pattern of the previous frame of the video. All patterns of each frame change their details "jump." Recall that the time between frames was 40 ms. Here we should especially note that all the processes that controlled the size of the pattern immediately before the beginning of the displacement had characteristic relaxation times of 200–300 ms, which significantly exceeded the actual time of a cardinal change of the pattern itself.

Another key result is that the axis of the output optical beam coincides with the axis of the input optical beam with all the "manipulations" with the cuvette with a colloid: its movement along the Z axis (±49z$_0$); micro-shaking due to the unevenness of painting the surface of the table on which the table with the cuvette was moving. Here we note that the cuvette was oriented at a small angle to the axis of the

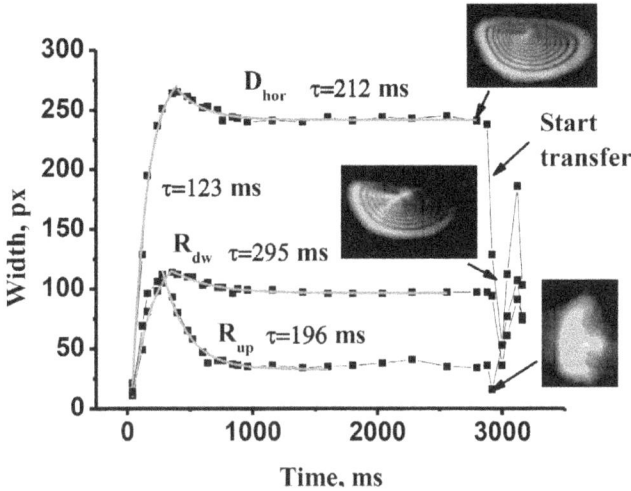

Figure 2.
The establishment of a stationary beam trace profile pattern. The inserts show the direct transformation of the pattern after the beginning of the movement from the position along the axis Z = −15z₀ to the side closer to the waist of the focused beam.

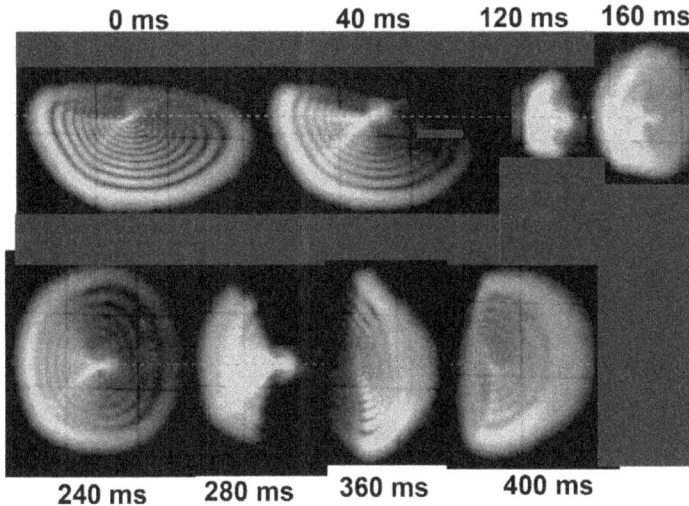

Figure 3.
Transformation of the pattern of the beam trace profile during the movement of the colloid from the position z = −15z₀.

input optical beam, and the axis of direct movement of the cuvette did not coincide with the axis of the input optical beam.

The pattern of the beam trace profile changes its structure and dimensions "abruptly" in each frame of the video. **Figure 4** shows how the digital profile of the beam trace profile pattern changes its structure and size after the beginning of the displacement (0 ms) of the cuvette along the z axis and after 120 ms. Here we have to remind that the beginning of movement took place after 3 seconds of continuous illumination, when the pattern of the beam trace profile was in a steady state with a characteristic exponential relaxation time $\tau \sim 200$–300 ms.

Figures 2–4 contain information that shows that the micro shake of a QDs colloid transforms the pattern of the beam trace profile over a time that is significantly

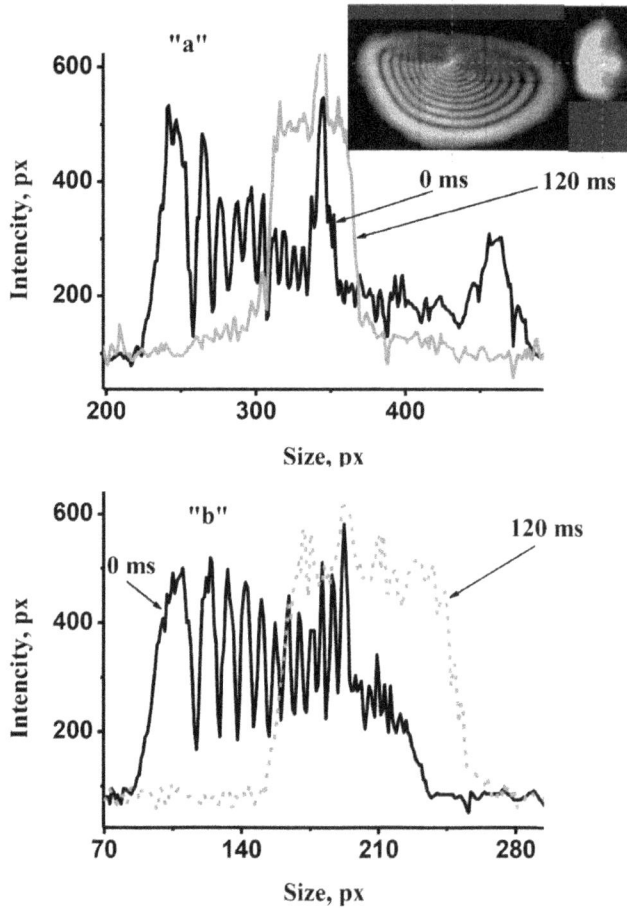

Figure 4.
Digital profile of horizontal slice "a" and vertical slice "b" of the beam trace pattern 120 ms after the beginning of the movement (0 ms).

shorter than the characteristic exponential relaxation time of the steady state of the QDs colloid. Here we recall that the pattern of the beam trace profile is a pattern of wave aberrations of the wave surface of the light-induced refractive index volume [17, 18]. The photoinduced refractive index of a colloid of QDs results from the accumulation of the concentration of QDs with a light-induced metastable exciton [16]. Consequently, the transformation of the pattern of the beam trace profile is the result of the transformation of the distribution of the concentration of QDs with a metastable exciton in the illuminated volume of the QDs suspension. **Figure 4** convincingly shows that a substantial concentration of QDs with a metastable exciton, providing phase addition to the wave front of the input beam, for example, at 14π disappears without a trace for a time shorter than the characteristic relaxation time of the steady-state stationary concentration of QDs.

In principle, this behavior of the QDs concentration is expected. Micro-shaking is a source of forces that can cause flows in a liquid, which mix the concentration of QDs. But, the fact is that micro-shock causes forces with an arbitrary direction. It is obvious that such forces should cause arbitrary concentration flows in a liquid, which should cause an arbitrary geometric displacement of the optical beam, its axis, in the first place. The experiment shows that arbitrary QDs concentration fluxes with a metastable exciton really arise, but all these fluxes "spin" around the axis of the input optical beam. The axis of the input beam has "unshakable" directions and retains its direction for all mechanical perturbations of the cell with QDs colloid. This means only one thing: there are no real flows of QDs concentration in the liquid, and what we see is the result of teleportation of the quantum states of a metastable exciton. Quantum teleportation "transfers" only quantum states from one quantum object to another quantum object. The trajectory of the transfer, of course, is absent. We have implemented a unique situation where mechanical classical forces are small enough to cause a real disturbance of the fluid, but these forces easily cause quantum teleportation, which does not have a trajectory of movement in classical space. The lack of a trajectory of movement clearly means that there is no actual movement of objects in space. Obviously, there is no movement; therefore, there are no forces that prevent this movement. This means that what we see is the result of the direct action of the forces not "burdened" by the opposition of any other forces.

The fundamental and practical significance, as well as novelty, of these results cannot be overestimated. The fundamental significance and novelty lies in the fact that the resource of entangled quantum states creates a macroscopic wave function regardless of temperature. Quantum teleportation transports quantum states of neutral particles, for example, quantum dots with a metastable exciton, without a specific trajectory of motion in classical space. Since there is no movement trajectory, then there is no movement itself. Movement is not, means that there are no forces that impede movement. There are no such force, which means that there is no internal friction. There is no internal friction in a fluid, for example, in a colloid of quantum dots, and there is a real displacement of a quantum dot, since a quantum state with a metastable exciton is another stable quantum state of quantum dots in classical space, and therefore it is another quantum object. Moving quantum objects in a liquid without internal friction is the basis for the implementation of a super-fluid quantum liquid, regardless of temperature. Superconductivity can be realized regardless of the temperature on the same quantum entanglement resource, but for this it is necessary to confuse the quantum states of charged qubits.

Practical significance and novelty lies in the fact that quantum teleportation allows you to register super-weak forces. Obviously, a super-weak force can impart to a super-small mass a sufficiently large acceleration, which is easy to register, especially in the absence of internal friction. On this basis, the possibility of developing super sensitive sensors, for example, for registration of gravitational waves, but in the size of an ordinary laboratory table opens.

To conclude this section, we formulate the physics of the quantum teleportation process of entangled quantum states. The obvious condition of quantum teleportation is that entangled quantum states must occupy a macroscopic volume. It is the volume in which the geometric displacement of quantum states takes place. In this work, this volume determines the geometry of the input optical beam, as well as, for example, in [20]. This optical beam light induces a second stable quantum state (QD with a metastable exciton) from the first state (QD in the ground quantum state), in other words, the optical beam generates classical two-level qubits, which at a sufficiently high concentration self-organize into a quantum state of quantum superposition with 2^N entangled quantum states. Decoherence takes place under the influence of both internal and external forces. It is under the action of these forces that the "disentangling" of 2^N quantum states into one of the stable states $|0\rangle$ or $|1\rangle$ of each individual qubit from N classical qubits takes place. The concentration distribution of these particular qubits is easily measured, since they are already in the classical space. The specific geometrical place where the quantum states $|0\rangle$ or $|1\rangle$ fall into is determined by internal forces (concentration compression as QDs accumulate with a metastable exciton) or external forces (whistle of the beam trace profile pattern). An analogue of the physics of such teleportation is the precipita-tion of raindrops (quantum states) from a macroscopic rain cloud (quantum superposition) under the action of internal forces (for example, the turbulent distribution of condensation centers) or external forces (for example, turbulent flows or wind gusts).

3. Quantum gravity as macroscopic force

Quantum gravity is a well-established term in the framework of the creation of the unified field theory, and this term means a quantum description of gravitational interaction. Obviously, the process of describing the gravitational interaction is not related to the emergence of gravitational force, as a fundamental force that plays a key role in nature. I propose to return quantum gravity to its original meaning as the primary source of interaction forces in nature. Quantum mechanics and general relativity are two fundamental theories that underlie the theory of quantum gravity. But, these theories are based on supposedly different a conceptual principle, which does not allow creating a unified field theory based on the theory of quantum grav-ity. Direct experiments in the field of quantum gravity are inaccessible to modern technologies due to the weakness of gravitational interactions. This is only a short list of difficulties that arise when trying to understand what quantum gravity is. I propose to combine the supposedly different conceptual principles of quantum mechanics and the general theory of relativity not to create a theory of Kant's grav-ity, but for the experimental realization of quantum gravity as a macroscopic force.

Obviously, any quantum object has mass. Then, the gravitational interaction between these objects, as bodies having a certain mass, is called quantum gravity by analogy with the classical concept of gravity. The mass of quantum objects is very small, and then it is obvious that the force of such quantum gravity, due to mass, can play a significant role and be detected at very small distances. It is believed that this distance determines the absolute unit of Planck's length, which is 10^{-33} cm. Penetration into the scale of units of length and Planck's time requires the creation of a density of 10^{99} cm^{-3} objects. For this you need to build a collider size, probably from the Milky Way. These are supposedly obvious direct experiments, the technol-ogy of which cannot be realized at the present time.

I propose another technology for creating quantum gravity as a macroscopic force precisely on the basis of the conceptual compatibility of quantum mechanics with the general theory of relativity. For this, I propose to assume that quantum gravity, as a force, is the result of the space-time curvature of the field from the point of view of the general theory of relativity. And from the point of view of quantum mechanics, the source of quantum gravity should bend the space-time field at the quantum level. The moving mass creates a curvature of the space-time field in classical space; therefore it is the source of classical gravity. Then, by anal-ogy, quantum gravity should arise as a result of a quantum process that bends the space-time field at the quantum level. Such a process exists and is generally known. This process is the transition of any quantum object from one quantum state to another quantum state, since the wave function of any quantum state transforms itself in space and time, and therefore bends space-time, with any change of quantum states. This is a well-known and generally accepted experimental fact. It is obvious that the curvature of the space-time field and, therefore, quantum gravity, as a force, will increase with an increase in the number of such quantum transitions. I propose to consider a quantum state called quantum superposition as a kind of quantum space that contains 2^N entangled quantum states, where N is the number of quantum objects that participate in quantum superposition. The reason for this proposal is simple. Only quantum superposition provides an exponential increase in the number of quantum states and, thus, entangles all quantum states of quantum superposition and provides an exponential increase in the number of transitions of quantum objects from one state to another as a result of decoherence of quantum superposition. Indeed, quantum superposition, for example, from $N = 1000$ quantum objects contains $2^{1000} = 10^{301}$ entangled quantum states, which become $N = 10^3$ stable quantum states in the classical space during the collapse (decoherence) of quantum superposition. Therefore, decoherence of quantum superposition of 1000 objects provides 10^{298} mutual transitions of quantum states, which is many orders of magnitude greater than the density of quantum states necessary for experimental work on the scale of absolute Planck units.

3.1 The results of the experiment and discussion

The experimental results of part 2 of this article substantiate the teleportation of quantum dots with a metastable exciton under the action of external classical forces. This teleportation is the result of quantum teleportation of the "metastable exciton" quantum state. This result looks like a fantasy, but this result is a physical reality, since the qubit is a quantum object in two stable basic states. This means that a qubit in the state (QD in the ground state) and a qubit in the state (QD with a metastable exciton) are different quantum objects in the classical space. **Figure 1b** shows the transformation of the pattern of the beam trace profile in the process of achieving a steady state. A nonlinear optical response is formed as a result of a photoinduced change in the refractive index. QDs with metastable excitons are the direct source of this photoinduced refractive index. Therefore, the stationary state of the nonlinear response is established when the concentration of quantum dots with a metastable exciton is established in the stationary state. And this is due to the accumulation of quantum dots with a metastable exciton, as a state with a long relaxation time. Therefore, the unique flattening of the upper half of the beam profile pattern should be associated with the accumulation of QDs in the state. The experimental results of **Figure 5** confirm this statement and show how the beam trace profile "comes" to its stationary state when the colloid was in the position $z = -25.5$ cm $= -49z_0$. The input optical beam had a radius $w = 1374$ μm, which causes sufficiently long diffusion times for the accumulation of quantum dots with a metastable exciton. This makes it possible to record all stages of the transformation of the beam trace profile in sufficient detail, since the registration took place with a digital camera with an interval between frames of 40 ms. Another "highlight" of the experiment in this position along the Z axis is that the intensity of the input optical beam was ~1 W cm^{-2}.

The fact is that the authors of almost all works consider that if the optical medium absorbs optical radiation, then the nonlinear optical response is thermal nonlinearity. Thermal nonlinearity is a consequence of a decrease in the density of the optical medium as a result of its heating. The non-linear thermal lens is, as a rule, negative and it defocuses the optical beam. The defocusing of the optical beam manifests itself as an increase in the size of the beam trace profile on a remote screen.

Figure 5 shows, with all the evidence, that the size of the beam trace profile coincides with the size of the input beam trace profile during the entire time of establishment of the steady state. This means that there is no thermal nonlinear lens, and a unique transformation of the output beam profile pattern is present. Therefore, the flattening of the beam trace pattern is a result of the action of forces that increase with the accumulation of QDs with a metastable exciton.

The obvious direction of action of these forces is shown in **Figure 6**, which shows the transformation of the pattern of the beam trace profile in the position of a cell

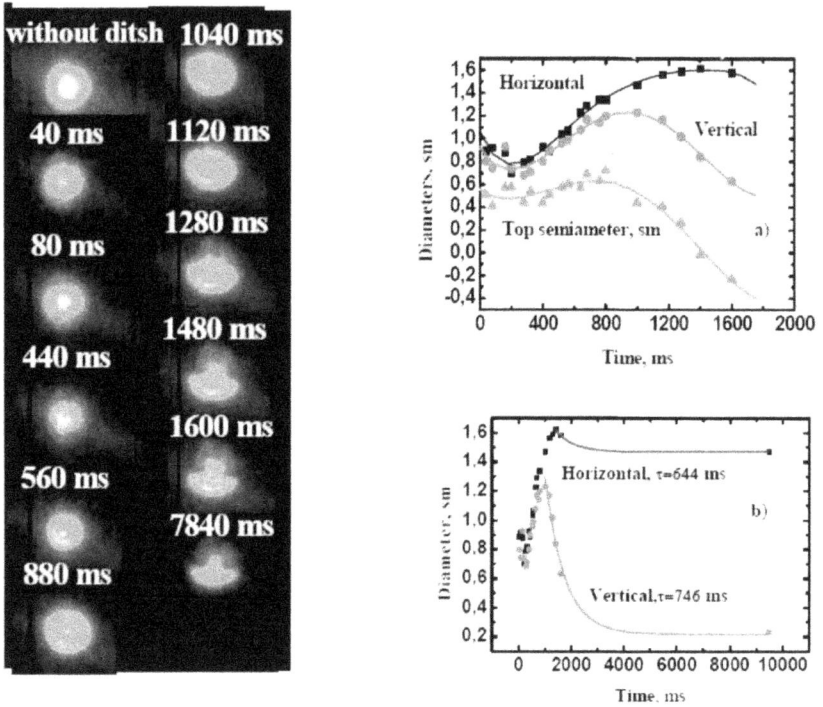

Figure 5.
The photographs represent the profile of the beam trace in the process of establishing a steady state. The numbers have time after the start of the lighting. (a) Represents the beginning of the development of transformation; (b) shows the relaxation of the beam trace size to the stationary mode.

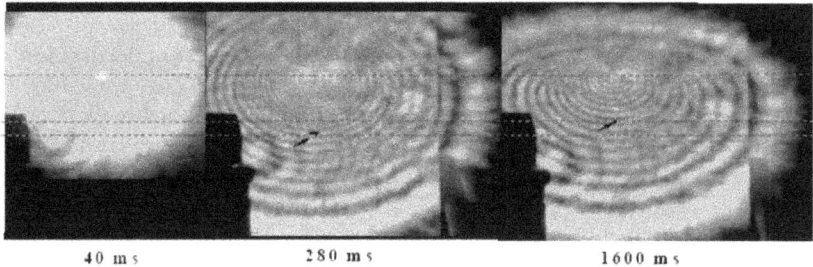

40 m s   280 m s   1600 m s

Figure 6.
Transformation of the pattern of the beam traces profile when the cell was in the waist of the input beam. The numbers have time after the start of the lighting.

with a colloid near the waist of the input optical beam. The input beam has a maximum intensity, and it illuminates the minimum volume of the nonlinear medium in this position. Therefore, the curvature of the wave front of the light-induced lens increases in comparison with the curvature at positions far from the waist of the input beam. The optical power of this lens also increases. The size and number of rings of the beam trace profile pattern increases, and the time to steady state is reduced. The files videos 1–3.gif demonstrates the transformation of the beam trace profile for this position of a colloid cell in real time. Collapse (self-focusing) of the optical beam takes place at the very beginning of illumination of a nonlinear medium. A typical Townes profile [21] is formed in the first 40 ms after the start of illumination. A dozen rings are formed already to 120 ms after the start of illumination. The increase in the number of rings and the simultaneous "lowering" of the whole pattern of the beam trace profile downward is observed in the time interval 160–600 ms after the start of illumination. Subsequently, the upper half of the beam trace continues to descend, forming only three contrasting rings that do not "go" beyond the horizon, as at $z = -48z_0$ and the rings of the pattern of the lower half of the beam trace profile "tighten" to their center, which is located on the axis of the input optical beam. As an example, the ninth from the outer ring, marked by a dot in **Figure 6** (280 ms), is shifted to the place of the 11th ring (1600 ms) during the time interval of 280–1600 ms after the start of illumination. Thus, we see that the upper half of the beam trace profile descends almost to the axis of the input beam, while the lower half of the interference pattern descends first and then, after some time, "tightens" up to the axis of the input beam. This is one of the key results of the experiment, and it indicates that the light-induced force is directed to the center of the input optical beam, i.e., it is directed to the axis of the optical beam.

The photos in **Figure 7** demonstrate the pattern of the beam trace profile when the colloid shines through in the vertical "bottom-up" direction. It can be seen that the beam trace profile remains axisymmetric all the time. Transformation of different parts of the beam trace profile is absent.

Obviously, the horizontal scanning of the colloid differs from the vertical in that the gravitational force of the Earth is directed perpendicular to the beam axis, whereas with vertical scanning the gravity force is parallel to the beam axis. In other words, we are in a situation where two mechanical forces have different directions. One force is terrestrial gravity, and it is directed vertically downwards, and the other force is light-induced force and it is directed to the axis of the optical beam. Then, the resultant force is the sum of two forces in the upper half of the beam trace profile, and there is the difference of these forces in the lower half of the beam trace profile. As a result, the upper half of the beam profile is compressed almost completely, and the lower half of the profile is compressed slightly.

Section 2.2 of this paper justifies the property of QDs to form quantum superposition with 2^N entangled quantum states under CW-illumination by an optical beam. The continuous repeating cycle "self-assembling quantum superposition— decoherence of quantum superposition—and self-assembling again" provides an unimaginably large number of quantum transitions "N states—in 2^N quantum states—decoherence in N states." These quantum transitions provide, in turn, an unimaginable number of curvatures of the space-time field with quantum objects, which are QDs. And this is quantum gravity in the literal sense: quantum gravity is mechanical force. Judging by the results, for example, in **Figure 5**, the light-induced quantum gravity force somewhat exceeds the force of the earth's gravity, since the

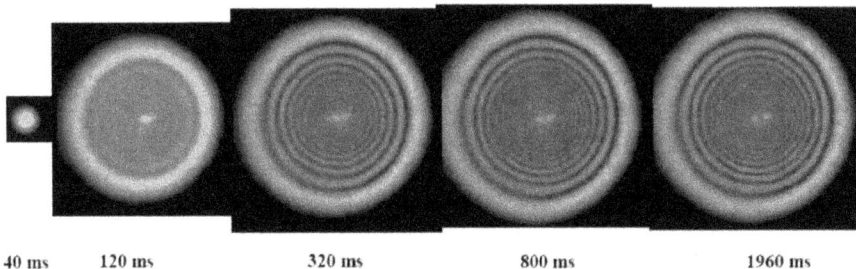

| 40 ms | 120 ms | 320 ms | 800 ms | 1960 ms |

Figure 7.
Profiles of the beam trace profile reflect the transformation of the progeny when the nonlinear medium shines through in the vertical direction: from bottom to top.

sum of these forces flattens the pattern of the upper half of the beam almost completely, and the difference of these forces slightly affects the pattern of the lower half of the beam trace profile.

The practical significance of such a force of quantum gravity solves the long-term problem of thermonuclear fusion of nuclei. The modern concept of nuclear synthesis suggests that plasma temperatures of 10^8–10^9 K will provide automatic synthesis of nuclei with a positive energy output. This concept is based on experimental results that are obtained repeatedly on particle accelerators. The real synthesis of nuclei in the H-bomb takes place, ostensibly, both because of the high temperature and because of the extremely high pressure, which arises as a result of the material being compressed by X-rays. X-ray radiation is an external force that can, in principle, compress the material, but this force is external and due to various kinds of fluctuations in the material, uniform compression is impossible, in principle. Quantum gravity is an internal force and, precisely, internal forces are capable of compressing the material evenly. Therefore, the real role of X-rays in the H-bomb is to create a quantum superposition of such a large N, that quantum gravity in the material of the H-bomb can be comparable to gravity in the center of the sun. The result is—lit a piece of the sun in terrestrial conditions.

4. Conclusion

In conclusion, the concept of quantum gravity was proposed for the first time as a force that arises as a result of the space-time field curvature at the quantum level when a quantum object passes from one state to another quantum state. The concept of real quantum space was first proposed as a quantum state of a multi-particle quantum superposition of N quantum objects, which contains 2^N entangled quantum states. The decoherence of quantum superposition takes these 2^N entangled quantum states into one of two stable quantum states for each of the N quantum objects. This process provides an exponential increase in the number of quantum transitions, and, thus, can provide conditions in which the force of quantum gravity reaches a macroscopic value.

Multi-level semiconductor quantum dots were first proposed as light-induced q-bits. The second stable quantum state of these q-bits is effectively separated from other excited quantum states of quantum dots due to the large relaxation time, which are six orders of magnitude longer than the relaxation time of other excited states of quantum dots. Light-induced charges carriers can be captured by surface-trapped QDs states and, thus, form an exciton with a long diffusion relaxation time. Such a quantum state of QDs is called QDs with a metastable exciton. Such QDs have a large individual additive to their refractive index due to the large spatial separation of charge carriers. The registration of the space-time redistribution of individual QDs was performed for the first time as the registration of the wave aberration pattern of the light-induced wavefront of the QDs suspension. This picture was the actual distribution of the refractive index in the volume of the QDs suspension illuminated by a laser beam. The experiments were performed under continuous illumination of the QDs suspension with a laser beam. This means that the quantum space (quantum superposition) functioned continuously according to the pattern of communication with the classical space: "creating a quantum superposition—decoherence of a quantum superposition—creating a quantum superposition again—decoherence of a quantum superposition again—and so on. Decoherence is the transition of entangled quantum states of quantum superposition into one of the stable quantum states in classical space. These two stable quantum states had different amounts of addition to their refractive index. It was these states that changed the light-induced wavefront in a QDs suspension, whose wave aberrations were recorded as a beam trace profile on a remote screen.

The experimental results show that two light-induced fundamental processes manifest themselves in the QDs suspension, which were revealed due to a significant change in the lighting conditions of the QDs suspension (Z-scan range from $z = -49z_0$ to $z = + 60z_0$) and time measurements, as the QDs suspension comes in its stationary state. The nonlinear optical response of the QDs suspension, as a nonlinear optical material, is the first process, the maximum value of which is achieved at $z = \sim z_0$. The emergence of a certain light-induced force, which flattens the upper half of the beam trace profile during horizontal scanning QDs suspension, is the second fundamental process. Such flattening of the beam trace profile manifests itself to the greatest extent under conditions when there is no nonlinear-optical response (very large detuning of Δz from the beam waist). The flattening of the beam trace profile is absent when the QDs scan suspension is vertically translucent, which suggests that the flattening of the beam trace profile has a gravitational basis. An analysis of all the details of the experiment allows us to conclude that the macroscopic force of quantum gravity was first implemented in the simplest laboratory conditions.

The result is fundamentally new; we can say there is a revolutionary one, not only for quantum physics, but also for the entire world view, from cosmology to the functioning of all life. Naturally, this result does not coincide with the modern concept of the development of quantum physics, especially if we take into account the complete absence of the mathematical apparatus for quantum space, therefore, the result is fundamentally new. The experimental results shown in this article, for example, quantum teleportation based on the formation of a macroscopic wave function, justify fantastic possibilities to realize the Bose-Einstein condensate regardless of temperature.

Author details

Anatolii Isaev
P.N. Lebedev Physical Institute (FIAN), Moscow, Russia

*Address all correspondence to: anaisaev@yandex.ru

IntechOpen

References

[1] Hoy J. Energetic and dynamics in quantum confined semiconductor nanostructures (Dissertation). Doctor of Philosophy in Chemistry. Washington University: St. Louis; 2013

[2] Sajad Y, Thompson PM. Nanoscale self-assembly of thermoelectric materials: A review of chemistry-based approaches. Nanotechnology. 2018;**29**(43):432001. DOI: 10.1088/1361-6528/aad673

[3] Stockert JC, Blázquez-Castro A. Chapter 18: Luminescent solid-state markers. In: Fluorescence Microscopy in Life Sciences. Sharjah: Bentham Science Publishers; 2017. pp. 606-641. ISBN 978-1-68108-519-7

[4] Gi-Hwan K, Pelayo García de Arquer F, et al. High-efficiency colloidal quantum dot photovoltaics via robust self-assembled monolayers. Nano Letters. 2015;**15**(11):7691-7696. DOI: 10.1021/acs.nanolett.5b03677

[5] Kwang-Tae P, Han-Jung K, et al. 13.2% efficiency Si nanowire/PEDOT:PSS hybrid solar cell using a transfer-imprinted Au mesh electrode. Scientific Reports. 2015;**5**:12093. Bibcode: 2015NatSR.512093P

[6] Chao X, Biao N, Longhui Z, et al. Core-shell heterojunction of silicon nanowire arrays and carbon quantum dots for photovoltaic devices and self-driven photodetectors. ACS Nano. 2014;**8**(4):4015-4022. DOI: 10.1021/nn501001j. PMID 24665986

[7] Quinn SD, Rafferty A, Dick E, et al. Surface charge control of quantum dot blinking. Journal of Physical Chemistry C. 2016;**120**:19487-19491

[8] Marchioro A. Recent advances in understanding delayed photoluminescence in colloidal semiconductor nanocrystals. Chimia. 2017;**71**:13-17

[9] Marchioro A, Whitham PJ, Knowles KE, Kathryn E, et al. Tunneling in the delayed luminescence of colloidal CdSe, Cu+-doped CdSe, and CuInS2 semiconductor nanocrystals, and relationship to blinking. Journal of Physical Chemistry C. 2016;**120**(47):27040-27049. DOI: 10.1021/acs.jpcc.6b08336

[10] Rabouw FT, Kamp M, van Dijk-Moes RJ, et al. Delayed exciton emission and its relation to blinking in CdSe quantum dots. Nano Letters. 2015;**15**:7718-7725

[11] Boyd R. Nonlinear Optics. Third ed. New York: Academic Press; 2007

[12] Horodecki R, Horodecki P, Horodecki M, Horodecki K. Quantum entanglement. Reviews of Modern Physics. 2009;**81**:865

[13] Schrödinger E. Die gegenwärtige Situation in der Quantenmechanik. Naturwissenschaften. 1935;**23**:807-812

[14] Lee YS. Self-Assembly and Nanotechnology. Hoboken, New Jersey: John Wiley & Sons, Inc; 2008

[15] Bokulich GJ. Philosophy of Quantum Information and Entanglement. Cambridge, England: Cambridge University Press; 2010

[16] Isaev AA. Two signs of superfluid liquid in a suspension of CdSe/ZnS quantum dots at room temperature. International Journal of Optics. 2019;**2019**:43638148

[17] Akhmanov SA, Krindach DP, Migulin AV, et al. Thermal self-actions of laser beams. IEEE Journal of Quantum Electronics. 1968;**QE-4**(10):568-575

[18] Whinneru JR et al. Thermal convection and spherical aberration distortion of laser beams in low-loss liquids. IEEE Journal of Quantum Electronics. 1967;**QE-3**:382

[19] Bennett CH, Brassard G, Crepeau C, et al. Teleporting an unknown quantum state via dual classical and Einstein-Podolsky-Rosen channels. Physical Review Letters. 1993;**70**:1895

[20] Ma XS et al. Quantum teleportation over 143 kilometres using active feed-forward. Nature. 2012;**489**:269-273

[21] Moll KD, Gaeta AL. Self-similar optical wave collapse: Observation of the Townes profile. Physical Review Letters. 2003;**90**:203902

Chapter 5

Quantum Dot Light-Emitting Diode: Structure, Mechanism, and Preparation

Ning Tu

Abstract

Quantum dot light-emitting diode (QLED) attracted much attention for the next generation of display due to its advantages in high color saturation, tunable color emission, and high stability. Compared with traditional LED display, QLED display has advantages in flexible and robust application, which makes wearable and stretchable display possible in the future. In addition, QLED display is a self-emissive display, in which light is generated by individual subpixel, each subpixel can be individually controlled. Each subpixel in LED display is constituted by liquid crystal and color filter, which make LED display have lower power efficiency and less enhanced functionality. This chapter introduces the QLED based on the QLED structure and light-emitting mechanism of QLED. Then, a novel method for fabricating QLEDs, which is based on the ZnO nanoparticles (NPs) incorporated into QD nanoparticles, will be introduced. The QLED device was fabricated by all-solution processes, which make the QLED fabrication process more flexible and more suitable for industrialization. What is more, as QLED devices were planned to integrate into a display, all-solution fabrication processes also make printing QLED display device possible in the near future.

Keywords: quantum dots, light-emitting diode, display, quantum dot light-emitting diode, material, printing

1. Introduction

After QLED was first published in 1994, a lot effort had been spent to improve the reliability and performance of QLED devices [1, 2]. The first QLED device uses thick QDs acting as both the emission layer and electron transport layer, which can be referred to in **Figure 1(a)** and **(b)**. The luminous efficiency of QLED device has been improved by the use of electron injection layer. Coe et al. published a sandwiched QLED structure in 2002, which consisted of two organic thin films with QDs as the emission layer. The luminescence was improved 25-fold over the best results of the previous QLED device, as shown in **Figure 1(d)** [3]. The QLEDs use Alq3 and poly-TPD as the electron transport layer, as shown in **Figure 1(c)**. However, the organic thin layers were sensitive to moisture and oxygen. Thus, replacing organic material with inorganic material seems to be the best solution to improve the reliability of the QLED device. Mueller et al. fabricated an all-inorganic

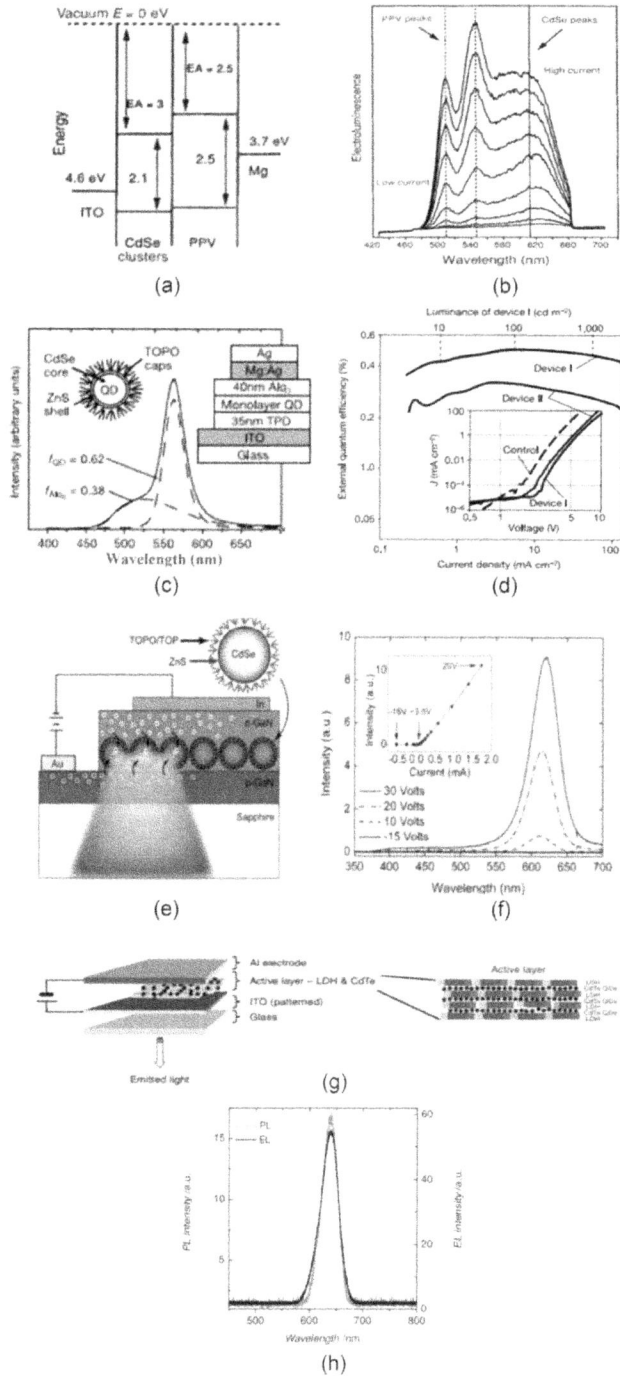

Figure 1.
Structure designs and material designs for efficient QLEDs. (a) Energy-level structure of the first QLED. (b) Electroluminescence of the first QLED [1]. (c) Electroluminescence spectra and structures for a 40-nm-thick film of Alq3, followed by a 75-nm-thick Mg:Ag cathode with a 50 nm Ag cap. (d) The corresponding external quantum efficiency [3]. (e) QLED structure consists of p-type GaN and n-type GaN. (f) Electroluminescence of QLED consists of p-type GaN and n-type GaN [4]. (g) QLED structure with metal oxide as the electron transport layer with layer-by-layer structure in emission layer. (h) Electroluminescence and photoluminescence of the QLED with metal oxide as the electron transport layer with layer-by-layer structure in emission layer [5].

QLED in which QDs were sandwiched by n-type GaN and p-type GaN. **Figure 1(e)** shows the structure of QLED with GaN, while **Figure 1(f)** shows the corresponding electrical performance. Mueller et al. used metal-organic chemical vapor deposition (MOCVD) method to deposit n-type GaN and p-type GaN [4]. However, the deposition method is too harsh for QDs that make the luminescence efficiency lower than the first type of QLED device. Then in 2010, Bendall et al. demonstrated an all-inorganic QLED with metal oxide as the electron transport layer. The QLED had a layer-by-layer structure, which had three-layer emission layers as shown in **Figure 1(g)** and (**h**) [5]. The overall device performance was still poor which is caused by the degradation of QDs during the harsh deposition process of inorganic materials. However, these devices showed condescending stability under long-term usage and high current density conditions.

The organic materials had an advantage of high luminescence, while inorganic materials had an advantage of high reliability. Then the researcher combined the advantage of both organic materials and inorganic materials by using both organic materials and inorganic materials as the electron transport layers. MoS_2, NiO, TiO_2, and ZnO have been reported as the inorganic charge transport layers (CTLs) [6–8].

According to the type of electron transport layers, the structure of QLEDs can be categorized into four different types (**Figure 2**): (a) organic/QD bilayer, (b) all-organic electron transport layer, (c) all-inorganic electron transport layer, and (d) organic/inorganic electron transport layer. The four different types of QLED structure also represent the development history of QLEDs in sequence.

Among these four types of QLED structure, inorganic materials are one of the most important choices for electron transport layers owing to their high electrical conductivity and good stability against environmental factors such as oxygen and moisture. ZnO nanoparticles (NPs) applied in electron transport layer are a significant breakthrough in QLED development history, due to their excellent electron

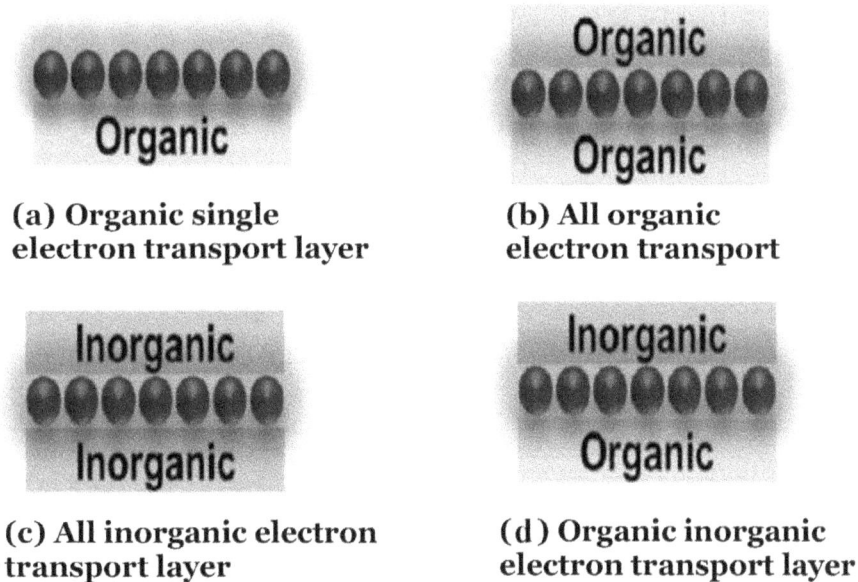

(a) Organic single electron transport layer

(b) All organic electron transport

(c) All inorganic electron transport layer

(d) Organic inorganic electron transport layer

Figure 2.
Four representative QLED structure types based on electron transport layers. (a) Organic polymer electron transport layer, (b) all-organic polymer electron transport layer, (c) all-inorganic electron transport layer, and (d) organic/inorganic electron transport layer [1–10].

mobility and no significant damage to the underlying QD layer during fabrication process. What's more, ZnO NPs are compatible with both polar solvent and nonpolar solvent, which makes the QLED fabrication process more flexible. More details about ZnO NPs will be introduced in Section 3.

2. Light emission mechanism of QLED

The emission mechanism of QLED is discussed in this subsection. A QLED has a similar structure and behavior as an OLED. In the QLED, the emitter is a semiconductor nanoparticle, while in the OLED, the emitter is an organic material.

2.1 Electron molecular orbital

Once a molecular orbital achieved the maximum electron energy, it is called the highest occupied molecular orbital (HOMO). Otherwise, if a molecular orbital has unfilled electrons, the molecular orbital is called the lowest unoccupied molecular orbital (LUMO). The energies of HOMO and LUMO affect the ionization potential and electron affinity of materials (**Figure 3**) [12].

Ionization potential energy is the minimum energy required to extract one electron from the HOMO, and electron affinity is the energy required to add one electron to the LUMO so that the system is stabilized [11, 12].

Before considering the light emission mechanism, it is important to understand the electron configuration in both the ground state and the excited state. Before excitation, when in ground state, the electrons are placed with both upward spin and downward spin (**Figure 3**). When excited, the electrons in the upper state are allocated with the same spin state, or the spin is reversed. The light emission is resulting from the energy transfer from the excitation state to the ground state.

2.2 Electron transfer and recombination

Normal materials in QLED have high resistance at weak electric fields. Therefore, researchers introduced the thin film to create strong electric field and chose

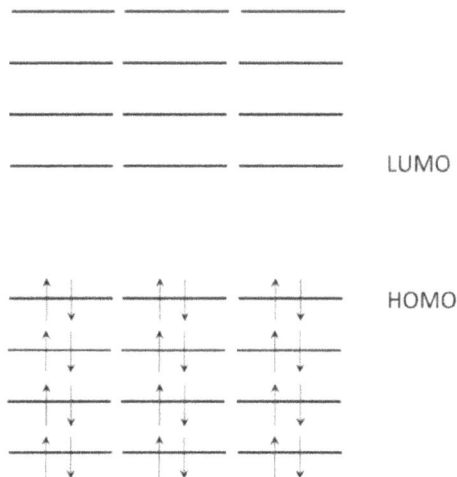

Figure 3.
The orientation of the HOMO and LUMO [12].

structures and materials suitable for charge injection [13]. The QLED performance is highly dependent on the choice of charge injection materials. Good charge injection materials should have high carrier mobility and balance the electron/hole injections well. The charge injection from electrodes follows the Schottky effect that means the injection barrier would be lowered according to the image force principle.

When an electron is injected into the electrode, if all the HOMO orbitals are occupied and cannot accept the additional charge, the charge will be transferred into the LUMO. When electrons are transferred into the LUMO, they form an electric current. At the same time, there will be a hole injected from the anode electrode which will be transferred into the HOMO. However, when the amount of injected charge exceeds the internal charge amount, the conduction system changes from ohmic to "space charge-limited current" [12–14].

If charge transfer by electric field and diffusion is taken into account and no trap is assumed, the electric current can be expressed as Eq. (1):

$$J = \frac{9}{8} \frac{\varepsilon \mu V^2}{d^3} \tag{1}$$

According to Eq. (1), the electric current is proportional to the square of the voltage. This is called the Mott-Gurney law, an extension of Child's law that takes collision into consideration [15].

The recombination and generation of excitons of QLED are shown in **Figure 4**.

When an electron and a hole recombined in the emission layer, the photons formed, whose wavelength corresponds to the energy bandgap of the quantum dots. The more electron and hole are recombined, the more photos will be generated, which corresponds to more light we could detect. Thus, people applied the hole transport layer (HTL) and electron transport layer (ETL) to restrict the electrons and holes in the emission layer, in order to improve the device efficiency. There are five typical layers for a QLED structure:

- Electron transport layer (ETL)—for electron injection from the cathode and transportation of the electron.

- Electron injection layer (EIL)—for electron injection from the cathode electrode.

- Hole transport layer (HTL)—for hole transportation from HIL to EML.

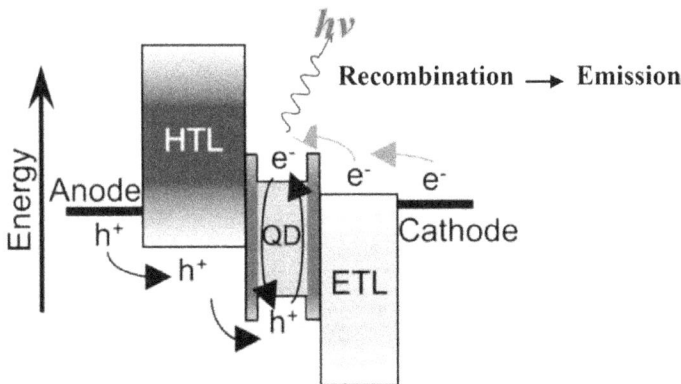

Figure 4.
The energy diagram of QLED.

- Hole injection layer (HIL)—for hole injection from the anode electrode.

- Emission layer (EML)—electron/hole transportation and their recombination to form an exciton; this is the QD layer in QLED.

This direct injection of charge carriers is assumed as the most common phenomenon for creating an exciton in the device.

In Section 1, the QLED structure types are elaborated. Thus, it is very important to design a QLED by factoring in the relationships between the work function of each layer. The QLED device fabrication process will be discussed in Section 3.

2.3 ZnO nanoparticles

Behind the QLED structure, the ZnO nanoparticles (ZnO NPs) have gained substantial interest in the research community as the charge transport layer (CTL). In 2008, Janssen [8] and co-workers demonstrated all-solution-processed multilayer QLEDs by using ZnO NPs as ETLs and organic materials as HTLs. The colloidal ZnO NPs were dispersed in isopropanol, and the deposition of the ZnO NPs on the top of the QD layers did not dissolve the underlying layers. Since then, continuous efforts were made to improve the performance of QLED with solution-processed n-type oxides as CTLs. ZnO NPs are widely used as CTLs in the state of the art of high-performance QLEDs.

Generally, solution-processed oxide CTLs can be deposited by two approaches, the precursor approach and the nanocrystal approach. The molar ratio of zinc precursor to potassium hydroxide (KOH) played an important role in determining the shape of ZnO NPs and hence affected the conductivity and mobility of ZnO NP film prepared from ZnO NPs [15–17]. ZnO NPs were synthesized by hydrolysis/condensation reactions under basic conditions. The synthesis procedure will be introduced in Section 3.

3. QLED fabrication by spin coating

3.1 Experimental chemicals

Detergent TFD4 was purchased from BioLab, PEDOT:PSS 4083 from Heraeus, poly-TPD (LT-N149) from Luminescence Technology Corp Ltd., patterned ITO glass from Xinyan Technology Ltd., green (CdZnSeS/ZnS) quantum dots from Suzhou Mesolight Inc., and zinc acetate dihydrate powder, potassium hydroxide flakes, acetone, isopropyl alcohol, methyl alcohol, chloroform, and chlorobenzene all from Aldrich.

3.2 Synthesis of ZnO nanoparticles

The uncleation-dissolution-recrystallization growth method [16] was applied to synthesize the ZnO nanoparticles. Firstly 0.37 g potassium hydroxide was dissolved in 16.25 ml methyl alcohol. Then zinc acetate dihydrate solution was prepared by adding 0.74 g zinc acetate dihydrate in 31.25 ml methyl alcohol at 60°C under vigorous stirring. Then the potassium hydroxide solution was jetted into the zinc acetate dihydrate solution at the rate of 0.8 ml/min. The reaction takes around 1.7 h under N_2 protection condition. After the reaction, the solution was allowed to sit for another 2 h to let the ZnO nanoparticles settle at the bottom of the reaction flask. The ZnO nanoparticles were washed twice by methyl alcohol. Then the ZnO nanoparticles were dispersed in chloroform and isopropyl alcohol mix solution at the concentration of 35 mg/ml.

Transmission electron microscopy (TEM, JEOL JEM 2010) was applied to characterize the morphology of the nanoparticles.

3.3 QLED device preparation

The patterned ITO glass was cleaned by detergent, methyl alcohol, acetone, and isopropyl alcohol in turn, each sonication for 20 minutes. After cleaning, UV ozone plasma was applied for surface energy modification. PEDOT:PSS was spin-coated on the cleaned ITO surface at 3000 rpm for 30 s and then baked under vacuum at 150°C for 30 min. Then hole transport layer (HTL) was prepared by spin coating poly-TPD on the annealed PEDOT:PSS surface at 3000 rpm for 60 s and baking at 100°C for 30 min under vacuum protection. Quantum dots were also deposited on the annealed poly-TPD surface by spin coating at 3000 rpm for 30 s. Then the synthesized ZnO nanoparticle solution was spin-coated at 1500 rpm for 60 s. The baking temperature is 60°C for 30 min. The cathode was deposited by vapor deposition method. A more detailed QLED device fabrication process was mentioned in previous work [2].

3.4 Results and discussion

3.4.1 ZnO nanoparticle analysis

The concentration of precursor, evaporation rate, and the time of reaction were all significant synthetic parameters, which affected the growth of ZnO nanoparticel dimensions and structures. **Figure 5** shows the TEM images of the as-synthesized

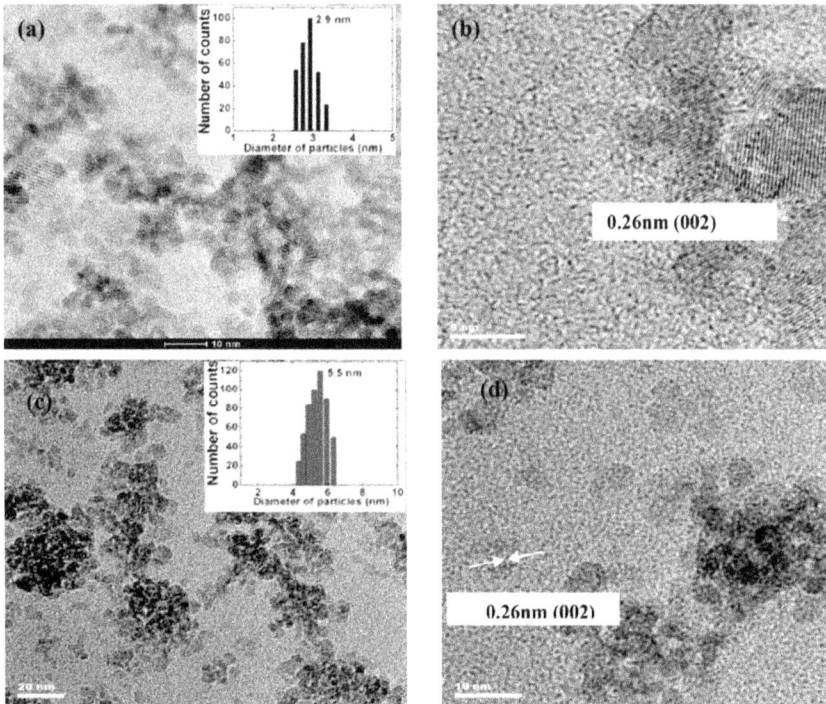

Figure 5.
TEM images of two different sizes of ZnO NPs. (a, b) reaction times of 105 min and (c, d) 80 min.

ZnO NPs from a well-dispersed ZnO colloidal solution carried out with the reference condition. To investigate the effect of reaction time on the growth morphology, reaction times of 80 and 105 min were carried out. The nanoparticle's size is smaller in the reaction of 105 min than in the reaction of 80 min. Moreover, the crystal lattice fringes are more clearly observed in the 105 min reaction sample rather than in the 80 min reaction sample. According to confinement effect, particles with smaller diameter would have higher energy. Therefore, the ZnO NPs used for QLED preparation are the smaller ZnO NPs.

In order to analyze the bandgap and quantum effects of the different ZnO NPs, their absorption and photoluminescence spectra need to be measured, which will be processed in future study. The energy bandgap (Eg) of the colloidal ZnO nanoparticles is determined from the intercept between the wavelength axis and the tangent to the linear section of the absorption band edge. The energy bandgap of ZnO NPs at 2.9 nm is 3.65 eV. The energy bandgap for the 5.5 nm ZnO NPs was 3.35 eV [17], while the energy bandgap of bulk ZnO is 3.2–3.3 eV [18], which is lower than the energy bandgap of ZnO NPs. It is found that the tendency of energy bandgap enlargement with decreasing size is consistent with the relationship based on effective mass approximation. Therefore, the reaction of 105 min can obtain smaller ZnO NPs than the ZnO NPs in the reaction of 80 min. In addition, the lattice fringes can be clearly observed in the TEM images, which suggests good crystallinity of the ZnO NPs.

3.4.2 QLED device performance and analysis

Figure 6(a) shows the structure of the QLED device, while **Figure 6(b)** shows the energy band diagram of the QLED device. The QLED device is a multilayer structure, which consists of PEDOT:PSS, poly-TPD, QDs, ZnO NPs, and Al. The thickness of each layer was measured by the surface profile (Alpha-Step 200 Tencor). **Figure 7** shows the TEM image of quantum dots; the diameter of the quantum dots was around 7 nm.

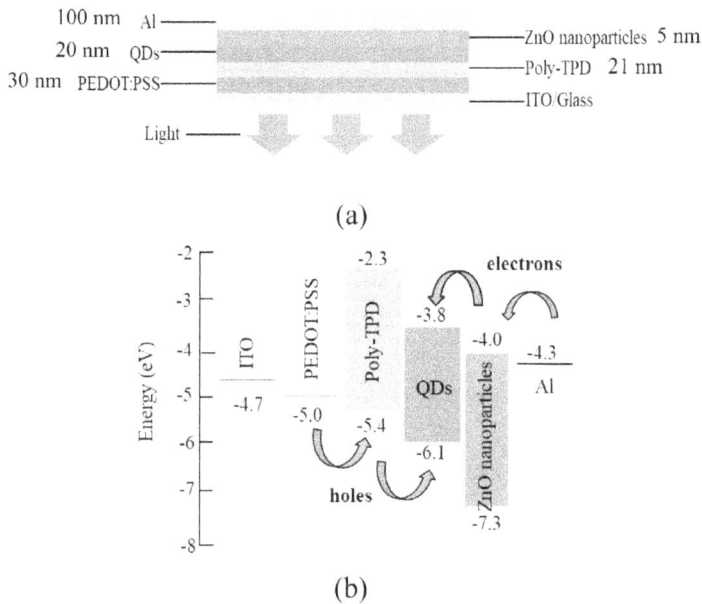

Figure 6.
(a) The QLED device structure and (b) the QLED device energy-level diagram.

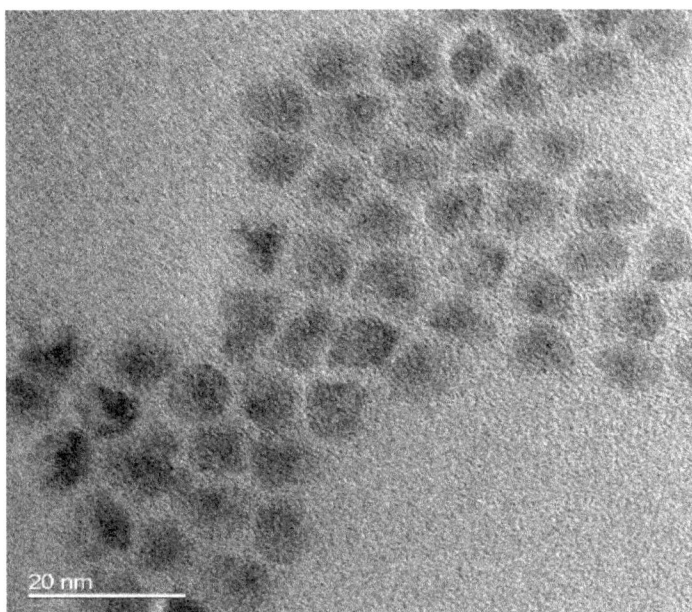

Figure 7.
The TEM image of green QDs from Mesolight.

The energy-level diagram in **Figure 6(b)** illustrated that the electrons and holes can be easily recombined together in the emission layer. Because ZnO NPs have a wide energy bandgap, the holes can be stored in the quantum dot layer. At the same time, poly-TPD's energy bandgap is from −2.3 eV to −5.4 eV. −2.3 eV is larger than −3.8 eV, which can also restrict electrons in the quantum dot layer. What is more, adding PEDOT: PSS layer and poly-TPD layer reduces the energy gap for holes jumping into the emission layer (quantum dot layer). The introduction of ZnO NPs has a similar function as PEDOT:PSS and poly-TPD, which confine the excitation and recombination region, hence potentially improving the efficiency of photon generation. The thin layer structure of QLED device makes it a promising candidate for the next generation of flexible displays.

Figure 8 shows the device performance analysis. The turn-on voltage is shown in **Figure 8(a)**, which is between 2 V and 3 V. The low turn-on voltage is due to the high electron mobility of the ZnO NPs and the design of the QLED structures. When there is a current applied to the device, the electrons can easily be injected into the emission layer, while the holes can also flow to the emission layer easily and be restricted by the ZnO NP layer. At the same time, the electrons accumulate at the interface of the poly-TPD/quantum dots due to the ~1.5 eV energy offset between the LUMO of poly-TPD and quantum dots. When one high-energy hole can be obtained after absorbing the energy released from the interfacial recombination of an electron/hole pair, the high-energy holes can cross the injection barrier at the poly-TPD interface and recombine with the electrons inside the QD layer and then emit photons. This is called the Auger-assisted hole injection [14, 19]. Therefore, the high electron mobility of ZnO nanoparticles and the band alignment structure can facilitate the hole transport and balance of the carrier injection of the device.

The electroluminance (EL) spectra of the QLED under different voltages are shown in **Figure 8(b)**. The inserted picture is the QLED device. The EL intensity of the QLED device increases as the applied voltage increases. The wavelength is 534 nm with the

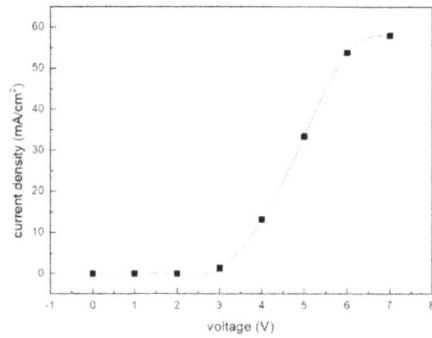

(a)

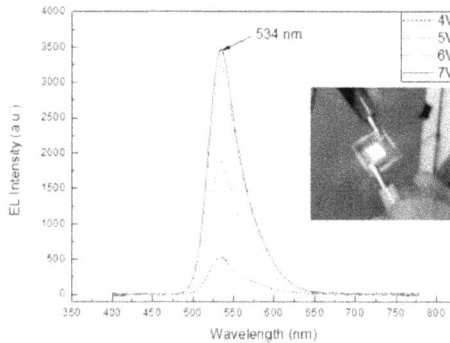

(b)

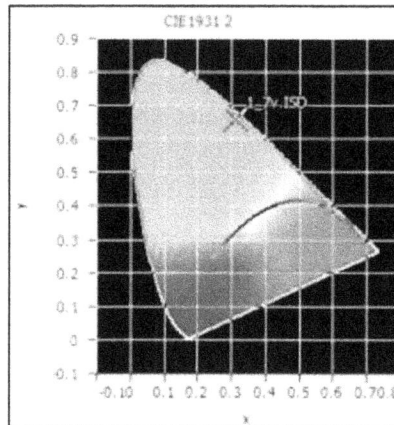

(c)

Figure 8.
(a) The QLED current density versus the voltage (J-V) curve, (b) the QLED electroluminance spectra as the applied voltage increased and (c) the 1931 CIE coordinate of QLED.

full width at half maximum (FWHM) value is 44 nm through the spectrum. The peak wavelength of QLED electroluminance spectra is 2 nm red shift compared with the peak wavelength of QD solution, which might be because of the dot-to-dot interactions in the close-packed solid film. Moreover, the electric field induced the Stark effect [20]. The 1931 CIE coordinate after emission is (0.31, 0.66) as shown in **Figure 8(c)**.

4. Conclusions

This work discussed the structure and mechanism of QLED and demonstrated an all-solution process of QLED in the last section. The QLED has high luminance with low turn-on voltage. These properties, caused by the use of ZnO NPs, improve electron injection and enhance radiative recombination. The resulting QLED fabrication process also makes printing QLED a possible method in the future.

Conflict of interest

The authors declare no conflict of interest.

Nomenclature

QLED	quantum dot light-emitting diode
QD	quantum dot
DOD	drop on demand
HDTV	high-definition television
CRT	cathode-ray tube
LCD	liquid crystal display
LED	light-emitting diode
OLED	organic light-emitting diode
PEDOT:PSS	poly(ethylenedioxythiophene):polystyrene sulphonate
ZnO NPs	zinc oxide nanoparticles
ITO	indium tin oxide
Poly-TPD	poly(N,N′-bis-4-butylphenyl-N,N′-bisphenyl)benzidine
EGBE	ethylene glycol butyl ether
ODA	octadecylamine
TOPO	trioctylphosphine oxide
DMSO	dimethyl sulfoxide
HTL	hole transport layer
HIL	hole injection layer
EIL	electron injection layer
ETL	electron transport layer
EML	emission layer
HOMO	highest occupied molecular orbital
LUMO	lowest unoccupied molecular orbital
λ	wavelength
h	Planck's constant h = 6.63×10^{-34} Js
P	particle momentum
E	the energy of a single particle in free space
m	mass of a single particle in free space
v	the velocity of a single particle in free space
k	the magnitude of the wave vector
L	standing wave/the diameter of QDs
J	current density
μ	carrier mobility of materials
V	voltage
ε	dielectric constant
d	the thickness of the thin film
FWHM	full width at half maximum

Author details

Ning Tu
Department of Mechanical and Aerospace Engineering, The Hong Kong University of Science and Technology, Kowloon, Hong Kong

*Address all correspondence to: ntu@connect.ust.hk; ningtu91@gmail.com

IntechOpen

References

[1] Colvin VL, Schlamp MC, Alivisatos AP. Lighting-emitting diodes made from cadmium selenide nanocrystals and a semiconducting polymer. Nature. 1994;**370**:354-357. DOI: 10.1038/370354a0

[2] Tu N, Kowk ZH, Lee SW. Quantum dot light emitting diodes based on ZnO nanoparticles. In: 20th International Conference on Electronic Materials and Packaging (EMAP). 2018. pp. 1-4

[3] Coe S, Woo WK, Bawendi M, Bulović V. Electroluminescence from single monolayers of nanocrystals in molecular organic devices. Nature. 2002;**420**:800-803. DOI: 10.1038/nature01217

[4] Mueller AH, Melissa AP, Marc A, Donald JW, Elshan AA, Daniel DK, et al. Multicolor light-emitting diodes based on semiconductor nanocrystals encapsulated in GaN charge injection layers. Nano Letters. 2005;**5**:1039-1044. DOI: 10.1021/nl050384x

[5] Bendall JS, Paderi M, Ghigliotti F, et al. Layer-by-layer all-inorganic quantum-dot-based LEDs: A simple procedure with robust performance. Advanced Functional Materials. 2010;**20**:3298-3302. DOI: 10.1002/adfm.201001191

[6] Caruge JM, Halpert JE, Bulovic V, Bawendi MG. NiO as an inorganic hole-transporting layer in quantum-dot light-emitting devices. Nano Letters. 2006;**6**:2991-2994. DOI: 10.1021/nl0623208

[7] Wood V, Panzer MJ, et al. Selection of metal oxide charge transport layers for colloidal quantum dot LEDs. ACS Nano. 2009;**3**:3581-3586. DOI: 10.1021/nn901074r

[8] Stouwdam JW, Janssen RAJ. Red, green, and blue quantum dot LEDs with solution processable ZnO nanocrystal electron injection layers. Journal of Materials Chemistry. 2008;**18**:1889-1894. DOI: 10.1039/b800028j

[9] Pan JY, Chen J, Huang QQ, Khan Q, Liu X, Tao Z, et al. Flexible quantum dot light emitting diodes based on ZnO nanoparticles. RSC Advances. 2015;**5**:82192-82198. DOI: 10.1039/c5ra10656g

[10] Choi MK, Yang J, Hygeon T, Kim D-H. Flexible quantum dot light-emitting diodes for next-generation displays. Nature Partner Journals. 2018;**10**:2-16. DOI: 10.1038/s41528-018-0023-3

[11] Bogue R. Quantum dots: A bright future for photonic nanosensors. Sensor Review. 2010;**30**:279-284. DOI: 10.1108/02602281011072143

[12] Tsujimura T. OLED Display: Fundamentals and Applications. Hoboken, New Jersey: John Wiley & Sons, Inc.; 2012. DOI: 10.1002/9781119187493

[13] Tang CW, Van Slyke SA. Organic electroluminescent diodes. In: IEEE/LEOS 1995 Digest of the LEOS Summer Topical Meetings. Flat Panel Display Technology. 1995. pp. 3-4

[14] Pankaj K, Jain SC, Vikram K. Current-voltage characteristics of an organic diode: Revisited. Synthetic Metals. 2007;**157**:905-909. DOI: 10.1016/j.synthmet.2007.08.021

[15] Ishii H, Kudo K, Nakayama T, Ueno N, editors. Electronic Processes in Organic Electronics: Bridging Nanostructure, Electronic States and Device Properties. 1st ed. Springer Series in Materials Science. Tokyo, Japan: Springer; 2015. 427 pp

[16] Seow ZLS, Wong ASW, Thavasi V, Jose R, Ramakrishna S, Ho GW. Controlled synthesis and application

of ZnO nanoparticles, nanorods and nanospheres in dye-sensitized solar cells. Nanotechnology. 2009;**20**:045604. DOI: 10.1088/0957-4484/20/4/045604

[17] Dai X, Deng Y, et al. Quantum-dot light-emitting diodes for large-area displays: Towards the dawn of commercialization. Advanced Materials. 2017;**29**:1607022. DOI: 10.1002/adma.201607022

[18] Ozgur U, Alivov YI, Liu C, Teke A, Reshchikov MA, Dogan S, et al. Comprehensive review of ZnO materials and devices. Journal of Applied Physics. 2005;**98**:041301. DOI: 10.1063/1.1992666

[19] Peng H, Jiang Y, Chen S. Efficient vacuum-free-processed quantum dot light-emitting diodes with printable liquid metal cathodes. Nanoscale. 2016;**8**: 17765-17773. DOI: 10.1039/c6nr05181b

[20] Kim HH, Park S, Son DI, Park C, Hwang DK, Choi WK. Inverted quantum dot light emitting diodes using polyethyleneimine ethoxylate modified ZnO. Scientific Reports. 2015;**5**:8968. DOI: 10.1038/srep08968

www.ingramcontent.com/pod-product-compliance
Lightning Source LLC
Chambersburg PA
CBHW081237190326
41458CB00016B/5820